Letters from East Germany

Letters from East Germany

The Postwar Journey of Christoph Haufe

LARRY ROTH

Westphalia Press
An Imprint of the Policy Studies Organization
Washington, DC
2024

Letters from East Germany:
The Postwar Journey of Christoph Haufe

Westphalia Press
An imprint of Policy Studies Organization
1367 Connecticut Avenue NW
Washington, D.C. 20036
info@ipsonet.org

ISBN: 978-1-63723-652-9

Daniel Gutierrez-Sandoval, Executive Director
PSO and Westphalia Press

Updated material and comments on this edition
can be found at the Westphalia Press website:
www.westphaliapress.org

Preface

I began this book with the intention of preserving primary source material for scholars researching what life was like for ordinary people in East Germany in the years after World War II. As I came to know Christoph Haufe (1925–92) through his letters, I realized he has a message for a wider audience. And as I researched his life after 1959, I found he made choices that enhanced his career, but which indicated a degree of ethical flexibility that some will find questionable.

Christoph refers twice to Margaret Mitchell's 1936 novel Gone with the Wind, once early on when he says he read a German translation before the war, not knowing he would witness devastation in Germany similar to what Mitchell described the American South as having experienced. Later, he is reading the book in English while he is recuperating from a severe bout of the flu. We can sympathize with Christoph, who was twenty-three when these letters begin, as he compares his life before the war, which included a pleasant home and a family car, to his life in a devastated Dresden, where his family can only heat one room, and when food is severely limited. The family goes everywhere not accessible by streetcar or train on foot or by bicycle, including his father, who is a minister, age fifty-two, and whose parish covers a large geographical area.

Mitchell's novel has fallen out of favor, and I doubt many people are familiar with 1,037-page behemoth. When I was growing up, my parents would often mention having seen the 1939 film, which was not re-released until 1967. I was in college at the time of the re-release, and a group of us went to see it on a wide screen. It was impressive, and it was a discussion topic among us for quite some time. We debated such weighty topics as whether Rhett killed Scarlett's second husband, Frank, when he rescued the vigilantes who wound up at Melanie's house. I was inspired to read the novel, which I found had lessons for the 1960s as the Vietnam War was raging. Among those lessons were Rhett's pointing out at the Twelve Oaks picnic that the North was in a much stronger position than the South—"there's not a cannon factory south of the Mason-Dixon Line … [there are] few iron foundries … or woolen mills or cotton factories or tanneries … not a single warship and that the Yankee fleet could bottle up our harbors in a week, so that we could not sell our cotton abroad…."

As was the case with Germany, it had been a generation since the last war, and between wars we tend to forget exactly what war means for those who fight it. At that same picnic, the elderly Mr. McRae warns those so eager to go to war, "You fire-eating young bucks, listen to me. You don't want to fight. I fought and I know. Went out in the Seminole War [1835–42] and was a big enough fool to go to the Mexican War [1846–48], too. You all don't know what war is. You think it's riding a pretty horse and having girls throw flowers at you and coming home a hero. Well, it ain't. No, sir! It's going hungry, and getting measles and pneumonia from sleeping in the wet. And if it ain't measles and pneumonia, it's your bowels. Yes sir, what war does to a man's bowels—dysentery and things like that—," at which point his family escorts him away from the crowd, since he's embarrassing them.

Finally, there's Rhett's response when he denies the war is sacred: "All wars are sacred to those who have to fight them. If the people who started wars didn't make them sacred, who would be foolish enough to fight? However, no matter what rallying cries the orators give to the idiots who fight, no matter what noble purposes they assign to wars, there is never but one reason for a war—money. All wars are, in reality, money squabbles, however few people ever realize it. Their ears are too full of bugles and drums and fine words from stay-at-home orators. Sometimes the rallying cry is 'Save the Tomb of Christ from the heathen!' Sometimes it's 'Down with Popery!' and sometimes 'Liberty!' and sometimes 'Cotton, Slavery and States' Rights!'" I imagine Christoph might have thought, and sometimes *Lebensraum*, racial purity, and all that came with Nazism as he looked at the ruins all around him.

As we witness from afar the devastation in Ukraine as well as Gaza, I think we should consider ourselves fortunate to have escaped such devastation in the United States—so far. As we risk both being pulled into an international conflict and the threat of a civil war here at home if national elections don't go one way or another, I hope we will take a deep breath and realize how lucky we have been, and how unlikely it is that we would escape such devastation in a future war. And I hope world leaders who keep us on the brink of international conflict and those who talk of a "new civil war" realize just what they would risk for themselves and their fellow citizens if either came to be.

In Christoph's letters we'll meet Emil Fuchs, who becomes very important to Christoph's development and career. Emil's son, Klaus, infamously leaked atomic secrets to the Soviets. My research on Emil led me to Nancy Thorndike Greenspan's 2020 *Atomic Spy: The Dark Lives of Klaus Fuchs,* which is a well-researched and quite readable biography that discusses the entire Fuchs family. The descriptions of Germany as Hitler and his thugs seized power is enough to make one wonder if it could happen here. The actions of the Nazis forced Klaus, who was an active member of the Socialist Democratic Party, to flee to Paris and then to England, where, after the war began, he would become an internee for a period of time as an enemy alien. Greenspan's description of the conditions the internees endured is almost unbelievable and include short rations and one shower a month! Klaus was interned first at Huyton, followed by the Isle of Man, and finally in Canada before being allowed to return to England, where he worked on England's atomic bomb project. He would eventually work on the Manhattan Project in Manhattan, Los Alamos, and Oak Ridge before returning to England. He was arrested on espionage charges on February 3, 1950. His arrest was front-page news around the world, and on February 9, 1950, an emboldened Republican Senator Joseph McCarthy gave his infamous speech in Wheeling, West Virginia. in which he claimed to have a list of 205 communists who worked in the U.S. State Department. Thus, Klaus Fuchs contributed to McCarthy's four-year reign of terror, which caused a great deal of paranoia and misery and ended the careers of many in academia and Hollywood, some of whom ended their own lives. An interview Klaus Fuchs had with the FBI led to the eventual arrest and 1953 execution of the Rosenbergs. Klaus Fuchs remained in prison until June 23, 1959. The day he was released, he was flown to East Berlin and settled in Dresden. Given Christoph's affiliation with Emil Fuchs, it's almost certain he met Klaus Fuchs as well as Emil's nephew, Klaus Fuchs-Kittowski, who became a student at the University of Leipzig in 1956.

In 1949, Emil Fuchs was living in Frankfurt, which was in the American Zone. The University of Leipzig offered him a position as a professor of Christian ethics and the sociology of religion. This was the position that brought Emil Fuchs into Christoph's life. It was also one of the stated reasons the British government investigated Klaus Fuchs, although by the time Emil took the position at Leipzig the investigation had been ongoing for quite some time.

The Letters

My father, Lawrence, died in April 2005. He was eighty-five years old, so, while his death was not surprising, it was unexpected. I had visited him the previous Christmas, and he was doing so well that when he wanted to go through what to expect when he died, which he did every time I visited. I told him he was in such great shape I shouldn't have to worry about that for a long time. Less than four months later, one day shy of the date on which my mother had died two years before, he was gone. April is indeed the cruelest month.

When my parents, Lawrence and Geneva, built their house in 1958, they included a lot of storage space. When my father died that space was full, as were three storage units in the back yard. My father had asked that we allow my brother, who, at the age of fifty-two, had essentially never left home, to remain in the house for six months. Eight years later he decided he'd prefer living in an apartment, so, in 2013, we put the family house up for sale and began the process of deciding what to keep and what to toss.

My father was born in 1920 in a dugout in Colorado, where my grandparents, immigrants from Europe, had homesteaded. In 1935, when the Dust Bowl was at its worst, they moved to Oklahoma City, where my grandfather, also Lawrence, was able to get a job on the streetcar lines. My father adapted well to city life, and in 1937 he was offered a scholarship in history at Oklahoma City University. My grandparents decided it would be better if he went to a business school, so they simply did not tell him about the scholarship. Let me interject here—I loved my grandmother, and it's not likely I would have survived my childhood without her, but she did have control issues.

As it became likely the country would go to war, my

father got a job with the Corps of Engineers in St. Louis, where he met my mother. When Pearl Harbor was bombed he rushed to enlist, only to be told he was medically unfit. He was very disappointed. The same thing happened to me during the Vietnam War, except for the rush to enlist and the disappointment. He was recruited to work on the Manhattan Project and moved to New York.

My father was not a physicist or any kind of scientist. He was merely one of about 130,000 people who worked on the Project. He worked in procurement, buying supplies for the Project. The Project must have really wanted him because the doctor who delivered him never filled out a birth certificate, and the Project had to reconstruct the events of January 1920. I suppose things were a bit informal in rural America at the time, and I'd guess this was so common the government had reconstructing birth certificates down to a science.

After a period in which he lived in New York and she lived in St. Louis, he and my mother married in February 1943. Their honeymoon consisted of a trip to New York, after which my mother returned alone via train to St. Louis. The Project hired my mother, and they moved to Santa Fe. Shortly after that move they went to Oak Ridge, where they remained "for the duration," as my father would say. They were then transferred to St. Louis, where they were let go. They moved to Oklahoma City, where my father worked as an accountant for a previous employer. He eventually became office manager. He remained there until the early 1980s, when he took the business over and moved it into his house.

Where did these letters come from? I honestly don't know. They were in his office files, so I assume that someone who visited the office may have given them to him. There were several things that fall into that category, including an unread copy of Joe McCarthy's 1952 *McCarthyism:*

The Fight for America accompanied by a pristine copy of the December 4, 1953 Chicago *Daily News.* I imagine he had to listen to quite a few rants in order to make the occasional sale. Another possibility, given that the recipients of these letters lived in North Carolina, is that someone he knew from Oak Ridge sent them. Christmas card management was a major project at our house at least into the late 1960s. (I moved to Columbus in 1971.) The entire family was enlisted to address and stamp hundreds of (fortunately) pre-printed cards. My father kept records of cards sent and cards received. (My system is if someone sends me a card, I send one back. Period.) Many of the cards sent and received were to people they'd known at Oak Ridge. Perhaps one of them knew Klaus Fuchs was Emil Fuchs' son and thought my father would find the letters interesting.

I'm very much interested in history, so I put several things aside to revisit. I'm ashamed to admit it was 2019 before I began reading these letters to see if I should keep them or toss them. It was not until Christoph's letter of February 13, 1955, in which he notes the tenth anniversary of the bombing of Dresden, that I realized these were indeed not only worth keeping but sharing. We get to know Christoph as he endures the poverty and deprivations of postwar East Germany, enters university in Leipzig, meets and affiliates with the renowned Emil Fuchs, and goes on to become a professor himself. We meet his family, see him through a failed love, and learn what life was like as postwar restrictions waxed and waned. And we learn about Christoph's life after the letters.

The letters end in 1959. I was very fortunate, while I was writing this book, to have had help from so many people. One of the most helpful was Dr. Phil Leask, a writer and researcher on German history and literature at University College London. He took the time to wade through a first draft and point out clues in Christoph's letters that hint at the path he chose. He also directed me to a website that allowed me to track mentions of Christoph in East German newspapers. I have noted the ones that occurred within the timeframe of the letters in bold and address the others in the last chapter.

The vast majority of articles that mentioned Christoph appeared in *Neue Zeit,* the official organ of the Christian Democratic Union of the German Democratic Republic. It and all media in East Germany was under the control of the government. It began publishing on July 22, 1945, and ceased publication on July 5, 1994. The Christian Democratic Union was founded in 1945 and was part of the National Front with the Socialist Unity Party of Germany. Unless indicated otherwise, the notes will be from *Neue Zeit.*

The Haufes: An Overview

As we'll see in the letters, Christoph Haufe's family consists of his father, Ehrenfried, his mother, Edith, and two brothers, Ehrenfried and John.

Christoph's father (1896–1964) makes a brief appearance in Doris L. Bergen's 1996 *Twisted Cross: The German Christian Movement in the Third Reich,* which discusses a little-known movement within German Protestantism that sought to expunge all Jewish elements from the Christian church. In a process that became more daring as Nazi plans for genocide unfolded, this group of Protestant lay people and clergy rejected the Old Testament, ousted people defined as non-Aryans from their congregations, denied the Jewish ancestry of Jesus, and removed Hebrew words like "Hallelujah" from hymns. Their services included preaching an essentially Germanic gospel and singing patriotic songs, including hymns of praise to the Führer, whom they placed on a level with Christ.

A goal among some of the German Christians was to unite all Christian churches, including the Catholic Church, into a national church. Bergen writes that many in the movement were opposed to this, including a "pastor in Dresden who lamented he was 'defenseless' when people discredited the movement by pointing to schemes for a national church and urged the movement's leaders to distance themselves from this 'supraconfessionalism'" (p. 112). A footnote reveals this minister to have been Ehrenfried Haufe (p. 264). This is the only mention of Haufe in the book, and he made this argument in October 1934, so while it's clear he identified as a German Christian in the early years of the Third Reich, it's not possible to know whether he remained part of the movement.

Christoph's failed love was with a girl whose mother was Jewish, and Christoph writes that his family concealed Jews. We know, of course, his father survived the war and the bombing of Dresden.

Christoph wrote a book titled *Die sittliche Rechtfertigungslehre des Paulus* (Pauls' Moral Doctrine of Justification) (Halle Niemeyer, 1957), which I've seen referenced in several other papers.

Christoph became a full professor in the Institute of New Testament Studies at the University of Leipzig in 1959, where he remained until 1984. I'll go into that in the penultimate chapter.

East Germany 1948–1959

I looked at several books on East Germany, and I found Katja Hoyer's 2023 *Beyond the Wall: A History of East Germany* (Basic Books) the most accessible in that it is arranged chronologically and told through the stories of individuals. Ms. Hoyer was born in East Germany in 1985 and is now a visiting research fellow at Kings College London. Other sources include Mary Fulbrook's 2005 *The People's State: East German Society from Hitler to Honecker* and a brief history of the East German uprising of 1953 issued by the U.S. State Department.

The following is a summary of the years 1948–1961.

The Marshall Plan, which provided economic assistance to restore the economic infrastructure of postwar Europe, began in 1948, but this aid did not go to East Germany. Instead, the Soviets, who wound up in control of East Germany, imposed reparations which amounted to confiscation of sixty percent of the country's production. Hoyer gives an example of this confiscation: In the early 1950s, Red Army soldiers would turn up at a tire factory and confiscate what had just been produced. The Berlin Blockade, which resulted in the Berlin Airlift, began just ten days after Christoph's first letter in June 1948.

In May 1952, Western powers signed the General (Bonn) and European Defense Community treaties. These treaties were a rejection of Stalin's March 1952 offer to agree to a unified Germany on the condition that it remained unarmed. The response to the rejection of Stalin's offer was a decision by East German Secretary Walter Ulbricht to unequivocally and quickly transform East Germany into a Soviet satellite state using a process called "Constructing Socialism." This involved forced collectivization of agriculture, a campaign against private trade and industry, and the development of heavy industry. The immediate result was severe food shortages, which began in the winter and spring of 1952-3. Forced remilitarization, the suppression of churches, and the persecution of opposition also added to the strain on the population. Christoph expressed concern that theological seminaries might be abolished in state universities in his letter of August 23, 1952. East Germans migrated *en masse* to West Germany, protest movements flourished, and expressions of general dissatisfaction characterized public life.

In April 1953, after the death of Stalin, Soviet authorities sought to rein in Ulbricht's regime, which was advocating even more collectivization. Ulbricht was persuaded to relax or discontinue parts of "Constructing Socialism," which was renamed the "New Course," but the treatment of manufacturing workers remained harsh. On June 16, 1953, a few workers called for a general strike, some throwing down their tools. Within days, nearly a million East Germans joined the protests and began rioting across hundreds of East German cities and towns. Soviet occupation forces declared martial law and used massive force to suppress the rioting and support the regime.

The U.S. for humanitarian reasons and undoubtedly also for propaganda purposes established a large-scale food relief program for East Germans, which began July 27. The U.S. pledged to distribute $15 million in food from thirty-five distribution centers in West Berlin, which was still accessible for East Germans through East Berlin. These "Eisenhower packages" contained lard, peas, flour, and pasteurized milk.

The distribution centers were able to distribute a little over five million packages to more than a million East Germans before the East German government cut off rail and bus service to West Berlin. I suspect this overreaction was one objective of the program.

Christoph's 1952 and 1953 letters reflect these events

as he constantly tries to keep his benefactors informed of what they can and cannot send. In one case, cocoa and coffee, which was extremely precious, were confiscated.

In 1955, West Germany joined NATO. Hoyer writes that Germans, east and west, were by and large tired of politics. She points out that a middle-aged German in 1949 had seen the whole spectrum of politics in their lifetime, none of which had provided a functioning democracy. The public was not interested in voting; what they wanted was food on the table, a restored roof over their heads, and a future that did not include war or economic disaster. The west was better at delivering these things. As the west improved, East Germany in 1952 moved families from the border areas, resulting in a ten-meter fortification patrolled by armed guards with orders to shoot trespassers.

At the same time, twenty percent of East Germany's potato crop was affected by a beetle blight, which was blamed on Americans who allegedly dropped the beetles from planes as a form of biological warfare. Food was scarce, as we'll see in Christoph's letters, and was rationed until 1958.

Hoyer writes that it took ten years to remove the rubble that resulted from the 1945 firebombing of Dresden before any rebuilding could begin, and the first new units were basic. Hoyer writes that a couple who were given priority for housing in Dresden because they were expecting a child were allotted a 388 square foot apartment with two tiny bedrooms, a small kitchen corner, and a small lounge. There was no refrigerator, no bathroom, and no running water. There was a tap in the central hall, which was shared by all residents. There was a public bath house. The couple lived in that flat eight years, and the wife, Hoyer reports, considered herself lucky because all the residents were in the same boat, and everyone supported each other.

To Hoyer's credit, she describes the bombing raids such as the one that did so much damage in Dresden as Anglo-American air raids. Too often I see Dresden used as shorthand for American atrocities. We can thank Kurt Vonnegut's first-hand description of the damage done by the February 13-15, 1945 bombings, which were indeed Anglo-American. Vonnegut was writing during the Vietnam War, which itself was an American atrocity. World War II was full of atrocities on all sides. I believe Dresden was in retaliation for the 1940 bombing of Coventry. A month after the bombing of Dresden, the U.S. firebombed Tokyo, killing at least 80,000 people—nearly four times the toll in Dresden. Hamburg was also firebombed, but neither Tokyo nor Hamburg had the good fortune to have their misery witnessed by a future best-selling author.

As the 1950s drew to a close, East Germany became more restrictive. While western music was not banned, radios and music venues had to play music from Soviet Bloc countries at least sixty percent of the time. Soviets were still demanding a good portion of East Germany's production. Farmers, including Christoph's older brother, were forced to join collectives as were skilled workers and tradespeople. Small businesses and tradespeople were forced into state-owned conglomerates. They lost both their autonomy and a good deal of their income. By 1961, East Germany had lost 1.3 million people, and the drain was projected to continue at a rate of 300,000 per year. Those who left were by and large intellectuals, professionals, skilled workers, and those who had owned large farms. More than eighty-five percent of the population left, hoping to raise their living standard.

On August 13, 1961, two years after Christoph's last note, the Berlin Wall went up putting a forced end to the exodus.

On to the letters.

The Letters

The letters are typed just as I found them. As we'll learn, Christoph is proficient enough in English to teach others, who pay him for lessons. I've corrected what I believe are typing errors, and standardized misspellings, grammar, etc., for the reader's ease and clarity. While in a few letters Christoph writes he is using a typewriter, I would imagine most of the letters were originally handwritten and the dates were written with the day preceding the month, as is the custom in Europe. In this first letter, we see that Christoph has received a package from America. I would guess he wrote a blind letter—perhaps to a church—in the hopes it would reach someone who would respond. I've done the same thing with some success. Before the 1989 revolution in Romania, I sent a blind letter to Sibiu, where my grandfather was born in 1894, asking if anyone knew anything about his family. After the revolution someone whose maiden name was Roth responded to my letter. I eventually went to Romania, and thanks to her, I was able to track that side of my family back to 1792. I suppose Christoph took the same approach I did—what did we have to lose by giving it a try?

Note that right off the bat, we find that Ms. Jameson has sent Christoph a return postage coupon he's not permitted to use.

Christoph never mentions how old he is. I initially estimated he was eighteen when the letters began; he was, in fact, twenty-three.

June 13, 1948

Dear Miss Jeannette Jamison:

It was quite a surprising joy when I received your letter the postman delivered yesterday. It's wonderful how fast airmailed letters travel and I am just sorry we aren't able to send out writings the same way because your coupon cannot be redeemed here but I will use it when I'll be in Berlin where I go sometimes in order to see some American shows. But nevertheless I hope it won't take too long until you will get my hearty thanks.

My joy was really unspeakable so I can't express how glad I am and can't find the right words to let you feel how happy I am my writing didn't take its way into some wastepaper basket but found people who understood me. Not long after I wrote that letter my courage and all my hope failed again because I thought it a big nonsense to write just into the wide world especially to people of a country where nobody knows me. Neither I could understand how I had dared to have written into a land which had such a lot of trouble and sacrifices because of the country I am a citizen of. Very often we Germans think we are despised by everyone and there will be an awful long time yet until these bad and terrible years will have gone by. But I was wrong, for my letter I sent over the ocean into a land which seems to us to be just like a paradise awoke some echo.

And I believe the greatest help is not the package I may get now but this helping hand of yours and this brotherly love which wiped out the borders of this world and gave me the assurance that we aren't forgotten in the mischief and misery which pushed us far below the normal level of mankind. This all I told you that you may understand how great my joy was when I read your letter and how thankful I am you wrote it.

Now let me introduce myself a little bit more. I think we will understand each other rather well since I am studying in order to become a teacher too. I am very fond of children, mostly of them who are about 12 years of age because at that age they seem to me to be most keen to acquire knowledges and experience and if the teacher is a skillful one he quick will win their hearts, but he has to be skillful because the age of 12 is also the age of impudence and they just look for the moment to cat-call the master of the class. This experience I got because I have to give some lessons in order to earn the money needed for studying.

Our university-building was a court-house in former times. But during the years of the last war a lot of buildings got destructed and turned into just a heap of debris, so also the university and high-school burned down. That's the reason the mayor of this city gave us some parts of the court-house. There we are quite comfortable. Every day I pass the big portal and attend my lessons and classes. I hear a lot about American culture, history and geography. So I can imagine quite clear where your home stands, what the surroundings and climate are like, what kind of life people live there, etc. Only I think one can't learn to know lands and people from books right but one has to get in touch with them, to talk to them, to listen to their stories, to read what they read, and to try to understand them. Maybe some day, years later, when times will be normal again and the choir of the Bethlehem angels will have turned out to have become quite realized, my wish will come true to see your country, and if you allow I will visit you too because I have to thank you so much. But quite some time I think, I must learn yet. About 2 years I have still to stay at Leipzig to finish my studies.

I lodge in a single and small room and have to keep house quite myself. So I live the life of a bachelor. On a little electric stove I cook and I turned out to become quite an artist in cooking. Well, one has got to, for it is really an art to cook nowadays, because one can't just go shopping and buying what the appetite is longing for. Very often my cupboard is empty and my ration-tickets all are spent and gone.

Now, when I will get your parcel, I think, cooking must be quite a fun. But now there is need on every line. When I go through the battered streets of my home city I always have to think about an American novel I read quite some time ago when I was a little schoolboy yet. Its title was *Gone with the Wind* by Mitchell. It was translated into German and read by many people at that time. All I remember now is there was a rich farmer with his family in the southern states. He was very prosperous and lived an extremely happy life. And then the civil-war came and he lost everything. He became poor and homeless, and his family members died. When I read the novel I was very touched but I didn't know the same fate was in front of my country. The time of my childhood was a very happy one.

My home-city was the most beautiful place of Saxony with famous buildings and treasures of art. Many foreigners liked to visit it. Upon the nice river which goes through it white-coloured ships sailed up and down carrying people into the near-by mountains with their old castles. My father, well, he was not rich, but he had saved some money, so we possessed a little car and a cosy home. And then the war came, and all this gladness was gone. People became very poor, missing the most-needed things for their life. Many of them lost everything, there is nearly no family of which no young member died during the years of this war. Our beautiful city was turned into just a heap of debris. I walk half an hour without seeing one house being in order yet.

Well, I believe that I will quit for this time, because it's getting late; but I will write again soon. Also I look forward to hear from you. Every day I will wait for the postman.

Sincerely your friend,

Christoph Haufe

A little more than a week later, Christoph receives another package.

June 21, 1948

Dear Miss Jeannette Jamison!

Today I received your well-packed parcel and it created an inexplicable joy. Since I had waited for it dayly I am awfully glad it didn't get lost. Oh it was such a happy feeling when I unpacked all those precious items you bought for me, and could store them in my empty cupboard. Now you may be sure cooking will be real fun and for some days there will be no trouble of being hungry and unable to work. Besides the contents were selected as expertly as if you were living here and knew what kind of food we miss most.

Each item is of such a precious value that I get sorry for having to eat it but would rather like to stand in front of it looking and wondering what a paradise-like country it must be where such kinds of cans are produced and such kind of food is available for everyone without rationing. We don't even get it on our ration points. So I feel like dancing around in my small room because of the happiness to be a possessor of riches seeming to be just scenery of a fairy-tale. But you may be assured I already stopped just looking and started eating since some inner voice urged me to do it and I wanted it to be quiet. Well, thus I followed its order.

Besides I got your box just in time because next week I have to deliver an address about the character-features of the American people. Therefore a lot of work has to be done and quite some books must be read. But besides I will certainly mention the fact of this overwhelming readyness of yours to help a person not being any relative to you. When I shall have delivered this speech I think there will be some time to write a detailed letter. Today I just want to inform you at once that your gift-package got to its destination, without any damage and made me able to forget about food-troubles just at a time I have to work so urgently because my address will get marked. I just wished I knew some way how I could express my thankfulness and give you some joy too. Well, perhaps I'll get an idea some day. Now at last I want to express my hope to hear from you and to get some letter that I may enjoy to have got some connection with the country of yours.

Yours sincerely,

Christoph

In this letter we see the ability of children to adapt to the worst of situations while, as Christoph writes, they "don't complain as the grown-ups do."

June 27, 1948

Dear Miss Jeannette Jamison!

Today we have a nice summer Sunday and if I wouldn't have to do some work yet I certainly would go for a swimming because the sun lures nearly irresistible, and swimming is my favorite sport. But the feeling of duty is a little bit stronger yet and so I have the best will to stay at home. If I'll succeed will turn out later on.

Dinner is just over and that is the reason I want to write you a few lines before I turn to my task again. Because today I opened the box of margarine you sent me and I prepared a dinner by which I got satisfied as I didn't since a long time. It was delicious. I baked some pan-cakes that tasted heavenly and the sweet and strong coffee I drank besides pushed my spirits on a level. I have to thank you so much for this package of yours. Presently I am in the right mood to sit on my desk at the window and gaze at the green tops of the green trees standing in the park nearby, and at the wonderful blue sky over them where some small white clouds wonder in western directions and in spite of knowing that they soon will vanish and disappear I dream like Peer Gynt they are beautiful wagons being able to carry me safely over the ocean Into the land of my dreams. But: "There's a long, long trail . . ."

In our class we presently are studying Milton's *Lost Paradise*. It's the story of the Satan who determines to wreck the happyness of men in order to take revenge on God. Satan enters paradise and in shape of a snake attempts Adam and Eve whose disobedience of God's order results in the loss of paradise. I always have to think it's an analogy of our present situation. The people of my home-country forgot to act according to the commandments of God and were therefore punished in a similar way. If I compare the present time with the years of the past, I am quite sure that we have lost a paradise, and if I read books of some American or English authors about every day's life in their countries I am even much surer that already in this transitory and passing world the areas are divided into heaven and hell. This thought is forced upon everyone walking through the battered streets of our cities where some years ago light-advertisements with big colored letters announced different shops, shows, theaters and articles. During night our streets are dark now. Only here and there some lonely street-lamps light the path and the entrances of some houses. One has to look very eagerly if one wants to see where the cinemas or the theaters are situated. Light advertisements aren't to be seen any more.

Also the shopping which was a delightful joy in former times is an ordeal nowadays. People don't enter the shops any more knowing that they can select various items from rich assortments. People walk through streets and where they see a long row of men they line up too and ask: "What can we buy here?" Children cry for a slice of bread or for a spoon of soup and not for candies, cookies or toys.

When I was a little boy, my father bought me a miniature train running by electricity. Nowadays most of the boys didn't even see such a kind of toy. They have to create their own toys or they paint some squares on the alleys over which they jump. Many haven't even a ball to play with. They play shopping and marrying, driving streetcar or riding aboard a train but they play it in a way different from the way children of your country do it. They are making games of the troubles around them. When they play shopping, they have small slips of paper which are presumed to be their ration tickets. Having entered the "supposed-to-be-shop" they ask if the 2 decade of bread is allowed to be sold, and the merchant tells them if there is or not.

Only a few of them have sound shoes. Many have to run bare-footed. They grow up and see the ruins around them. They play there and sometimes it even happens one is buried under a crashing wall. They sip a glass of colored water as we sipped the cup of cocoa they don't even know. But the miracle of all is: They are content. They don't complain as the grown-ups do (and as I did too). They smile and joke when playing and they listen to the stories of pre-war time just the same way we listened to the fairy-tale of *Hansel and Gretel.* But just the contentness of them makes me unspeakable sad and because they don't even know the paradise their parents lost on account of their own fault.

I have a two-year's old godchild, and I want that he will see some times. His second birthday is in autumn, and I would like to create him a real joy. May I ask you, therefore, to send me some cocoa perhaps, if possible. Well, I think you soon will get fed up with the sad thoughts I molest you with every time. But I guess it's the way we Germans think nowadays.

There is just one refugium to me and that is my studies. When scientifically working each other's problems seem to get more and more unimportant. It's some kind of a dream I live in.

"I wander over yonder till I see" the marvelous areas being not restricted and rationed but free to everyone who knows how to find the sphere of culture. But even that way will perhaps be taken away from me because—as you may have noticed from the lot of stamps on the envelope—the currency reform took place and nearly all my saved money is worthless by now.

How I will manage to go on studying I don't know yet. Certainly I have to take up some work besides, and it is not easy to find it. Besides the present food-conditions don't allow the mind and body to work nearly all hours of day and night. Many will have to give up their studying. But I didn't perish until now, and thus I am confident I will go on. May I hope to get some help from you? Don't get kept off by rumors about bad mail-deliveries because the Russians let pass every parcel coming from abroad. Difficulties are just in sending consignments from one zone of Germany to another one. Many people and neighbors get parcels from abroad regularly, and only seldom it happens that one gets lost.

But time is going on and so I will call it a day. I think I will go for swimming in spite of the work. I can do it tonight. I quit with the promise to write again very soon.

Yours sincerely,

Christoph

In this next letter we learn Christoph has a new benefactor. He details the truly meager rations Germans are given four years after Germany's defeat and dreams of life in a Shangri-La.

June 1, 1949

Dear Miss Grover!

Last week I received your letter of May 9 and was very glad to get it since I had already given up all my hope to hear from you again. It wasn't my main wish to get just a parcel of food, however much I may need it, but what I wanted was to get a connection with your country and its people.

I think everybody got his fairyland he dreams of sometimes. For some it is just an island of their fancy as perhaps in Hilton's *Lost Horizon,* where a beautiful land of peace is described being never touched by war, misery or any destruction of this world. No sickness is allowed to enter and the death has not the almighty power he has with us. I was very impressed by this novel of an American author you may know by chance. *[Hilton was actually British.]* It would be heavenly if the world would be like this. But since I know that this kind of living is promised to the disciples of God and that this very world we live in is like the examination for getting hold of a ticket for that place, I don't expect it could ever be realized in this world. What would we else need a heaven for?

My "Lost Horizon" is not such a fancy bit. It is a land which certainly may be glorified by my fancy, but nevertheless is a real and concrete one. It is the land Columbus set his foot on first, and Mr. Penn took as a refugium for the persecuted independents. It is the country where Jefferson's ten amendments of individual freedom were realized and all nations of this world were molded to a single one. Since years already I read each American book I can get hold of, and that may have been the way of me learning English as a language not only to be spoken by means of intellect and translating but as a language which has its own character and feelings. Perhaps it isn't in your country like I fancy it, but when I look at pictures and see the beautiful towns and landscapes (it aren't the big cities with their sky-scrapers I am mostly impressed of), and when I read stories about life there or listen to Radio-Frankfurt serving American forces in Europe I feel like sitting in a theater seeing happy human beings walking across the stage. It seems to us, sitting in the darkness, like another, like a glorified world of its own.

You ask me if I am still studying. Well, from the enclosed pictures you will see how fiercely I am occupied by it. Until next year I have to go on yet; then I will pass my examination. English isn't my leading topic, it is more a hobby. Old languages as Greek and Latin, besides History, Psychology, Philosophy and Religion I have mainly to work for.

About the conditions, if they are better?

Well, the crops were very good last year, but in spite of it we don't get more food. Just potatoes and vegetables are free yet. Perhaps you may be interested in the quantity and quality of food we get. Are you? Well, I will tell you about our rations but I ask you not to regard it as a whimsical or deploraable complaint but just as information. There are 5 groups of ration-cards with different rations. The best is the Number One. My father gets number 3, and my mother and brother number 4. The students get number 2. Number 4 is unofficially called the "starving ticket" but in spite of it nobody of my family starved yet. We are just hungry often, and that, as doctors say, isn't unhealthy. Isn't It? Number 4 gets about a half or a third of number 2.

Number 2 receives each day: 435 grams of black bread, 15 grams of white bread (that is: every four days, one slice of bread), 50 grams of noodles, oats, etc., 35 grams of sugar or sweets, 35 grams of meat or fish or sausage (but mostly we get only fish. Last month we received just 300 grams of meat), 15 grams of bones, 30 grams of fats (oil, margarine, butter), 30 grams of jam, 4 grams of coffee-surrogate. That is all and everything. There is no food without tickets besides potatoes and vegetables. There are no fruits (therefore, we are glad to have a garden), no cocoa, chocolate, no milk, no lemons and no lemon- or grape-powder, no coffee, no tea, no . . . Well, I could go on enumerating until tomorrow. You ask me what I need most. Well, it is easyly to be figured out. May I ask some cocoa and chocolate for Konrad. Naturally I like it, too, but I think for me it is necessary to have a useful standpoint. I would be glad to get some coffee. Sometime ago I heard that there was a sort of coffee which didn't leave any sediment but dissolved by itself. Is that true? But when going on to ask, I really have the feeling arising to check myself. Who am I that I claim anything? I will be more than glad for any help I get, but it is difficult to give you some kind of order. I know prices aren't low and each penny will be worked for. I frantically ponder how I could express my thankful feelings towards you. It is already for this reason terrible that we Germans became such a poor people.

Well, it's getting late and I barely can keep my senses together. I will try to write a better letter soon.

Sincerely yours, Christoph

Please, give my love to Miss Jamison.

Christoph remains very thankful for the gifts his American benefactor continues to send and has found a gift he hopes she will appreciate. I found his belief that "Americans have everything, don't need anything, are the richest and happiest people, can have everything their hearts long for" amusing. The first time I visited Romania, in 1990, I was told by friends I made there that America is a paradise. When I tried to explain that it was not a paradise for everyone, and that there are homeless people in America, they simply did not believe me.

June 13, 1949

Dear Miss Grover!

Three days ago I received your valuable gift-parcel. It got delivered quite undamaged and nothing missed. Oh, we all were very glad! Now we don't have to eat any dry bread for weeks. And there will be no lack of fats during the time; that will enable us to prepare meals we can't think of else. Our cooking has always to be done so terrible economically, no waste; and not even the needed things for good meals available. So we thank you very emphatically for your sacrificing such a lot of money. Also Konrad thanks you. When he got your gift, a real smile lighted his face. He has still to stay at the children-hospital. It will be a long time yet, the doctors say.

During the last time I often wondered and pondered how I could find something by which I'd be able to express you and Miss Jamison my thankful and loving feelings towards you. I asked nearly all my acquaintances if they knew what items the American people would be glad to get from Germany. The answer was always an amused smiling about my childish question.

Americans have everything, don't need anything, are the richest and happiest people, can have everything their hearts long for—in their own country. And what am I? A citizen of the destructed and vanquished Germany where people like to eat things Americans are accustomed to throw away. Where all the items having been famous all over the world formerly aren't available for common citizens anymore. And from this point we come to the next reason: Who am I? A poor student, without money to buy anything on the black market or in the shops owned by the government, where on might buy many things for high prices; a student, who is accustomed to sit on the cheapest seats when visiting shows, theaters or concerts. Well, you see the differences are too great as to find any way of compromise. I always will be debted to you.

But some day ago I got an idea how I eventually could give you some joy nevertheless. I was walking through a battered street when I saw the decoration of a small, scantily rebuilt shop. And by seeing it I remembered beautiful days of vacation I enjoyed together with my parents in the nearby mountains called: "Erzgebirge." It happened in times of peace but also during the first years of war that nearly each summer we travelled to a small village in this mountain region. Well, the mountains there aren't to be compared with the Rocky Mountains or the German Alps; they are softly sloping and are covered unto the tops of their hills with spare vegetation, in the valleys even potatoes and some kinds of grain are growing but very poorly. And therefore people living there were [for] hundreds of years already existing on a very low standard of life. Their huts were barely seen above the ground. In winter time snowstorms covered them with a white coat and carpet completely. At that time they sit in their small heated room, surrounded by the walls of snow and sing the songs of their homeland because in spite of the rough and raw treatment the weather imposed upon them and the spare harvests and crops often having been completely destructed by the coldness of an early arriving winter, they love their country with their most ardent hearts. In this part of Germany the most beautiful home-songs are composed, they speak some kind of a nicely sounding dialect reflecting all their warm feeling towards their home and so more and more travellers like to stay there during vacation-time and breathe this hearty atmosphere of plain and nature-bound people. In summer the wide forests with their healthy air and in winter the vast snow-landscapes really made for skiing allure them. Well, and that is a very good source of money for the inhabitants of the Erzgebirge.

But they have also found some other ways to keep themselves away from starving. During the long winter months, they don't sit around Just chatting and lazily sleeping, but far into the nights their weaver's looms are humming and their lace pillows are worked on until showing the nicest kinds of bobbin-work. But besides there is another skillfulness they developed and left from generation to generation. They carve.

Timber is the richness of their forests and with deft hands they made figures of all kinds and sizes from it. Grandpa and grandson sit together and emulate to produce the most impressing things. Houses, trees, dogs and every kind of animals, men and angels, articles useful for the household as spoons, twirling sticks, etc. But most famous they became for their toys. Little streetcars, trains, figures of all kinds, Christmas-Pyramids which are turning around driven by the heat of burning candles, wooden men with pipes and hollow bellies in which fumigating candles are placed, and hundreds of other wonderful creations. And these little figures I saw in the rebuilt shop I told you about. And as far as I know there seemed me to be nothing of that kind in America.

And so I thought: Perhaps you would get some pleasure from these beautiful figures made by the citizens of the Erzgebirge. I send two little packages, one to you and one to Miss Jamison. In one there are three Congratulation-Angels. The first one brings a message-letter and some flowers and the two others are (behind the first one) carrying numbers or figures that may be changed according to the anniversary concerned. In the other package you will find a group of children at a children-march. I selected different gifts since I thought you may exchange them according to your taste, or if you love all of them I will send the same ones a second time. Anyhow, if you love them, I would be glad too, and I would financially try to send you regularly. There are the most beautiful arrangements. (An angel-concert, etc.)

Perhaps there is some place for them in the technically established homes of you. They certainly will be astonished to find themselves among such surroundings but they will accommodate themselves and then they will start to tell of their lives, of the homely forests with their wonderful rustling, of the screaming saw and the wounding axe which cut them from the nourishing ground, they will tell about the carving-knives and about the poor, many-headed families sitting around the table while candles are spreading their fancy-light, and while snowstorms are roaring outside, but home-songs are being hummed through the warm little room combined with the crackling sound of the burning wood-sticks. Perhaps then they will get a little bit homesick when seeing themselves surrounded by all these American technic. And so I ask you to cherish them a little bit and to listen to their stories and their begging not to forget all about the country and its population they come from. They have their own kind of feelings; the sap drawn from the earth of their home is running through their veins.

I hope this little gift of mine will get safely through to you and will be a mean of understanding between our countries. If you get some pleasure of it, put it upon a place it will decorate your room from, so it may always see you, and you and your visitors will notice it and listen to their story. I hope they will be understood in spite of their foreign language since they come from Germany.

Sincerely your friend,

Christoph

In this letter we learn more about Christoph's family as well as his situation. He is a student living in Leipzig and visits his family in Dresden when he can. As what Americans can send to East Germany becomes increasingly limited, having two addresses will come in handy. As a student, he takes work where he can. He works on his brother's future in-laws' farm. He is fluent enough in English, Latin, and Greek to give private lessons in those languages, for which he is well paid (for the time, place, and circumstances). This fluency will serve him well; he will eventually become a professor of New Testament studies at his university. At this time, he is teaching colloquial English based on the novels in English in which he has immersed himself. He's not a fan of Hemingway, who admittedly is not everyone's cup of tea.

Evidently Ms. Grover has complained about American students' lack of discipline—in 1949! I suppose the everyone considers bygone eras "the good old days." Christoph advises her to look to the future rather than the past. I can't help but wonder how Ms. Grover would react if she were able to visit a modern classroom.

Christoph is sending some Erzgebirge figurines to America. Examples of Erzgebirge folk art can be found on the internet.

Konrad's story will become clear as time goes by. He is the half brother of the girl Christoph fell in love with. The romance didn't work out, but Christoph still does what he can for Konrad. Konrad is hospitalized for tuberculosis. Food was in such short supply after the war that it was rationed. According to Mary Fulbrook, the rations in 1946, when Konrad was infected with the disease, were about half what an average person would need. As a result, disease was rampant. That year there were, according to Fulbrook, 56,780 cases of infectious tuberculosis, of which 23,600 people, including Konrad's father, died.

August 13, 1949

Dear Miss Grover!

The theory of the mind being affected and motored by surroundings and impressions seemed very familiar to me when I received your letter a few hours ago. I got it just at a time I am rather worn out by the hard work of the summer-term, scarce food and other trouble. But the fatigue was overcome at once when your thoughtfulness surprised me. I was feeling really re-born when getting reassured the bond between us was a real strong one. The Lord may bless you for your kindness, and I am certain he will do it.

I enclose a picture of my family (my parents and my two brothers) that was taken when visiting an acquainted family (whose daughter is the fiancé of my brother Ehrenfried) who own a vegetable and flower garden. I worked there sometime. It was a dark time and the worst way of earning money I need for studying. At present I give some private lessons in English, Latin and Greek and enjoy them very much. Also my pupils don't seem to be troubled by them and make good progresses. So their parents are very content with me and pay me well off. They sometimes like to listen too.

I tell a lot of American stores and since I got a bit of slang-influence (because I prefer American authors) we have quite some fun. Naturally the schoolteachers mind this way of pronunciation, they don't judge upon it as "fine," but I can't think of any better, it seems me more hearty and good-natured than the high-nosed "King's English." it runs quite from the heart and I don't trouble to change it, even if I should get a worse mark.

Presently I read a novel by Booth Tarkington, *Little Orvie*. Each minute I can spare, even in jammed streetcars, I have to go on reading. It gives me such a lot of joy, that I would rather like to translate it. But that seems a lot of difficult since it isn't sufficient to translate only the words into the German language but the genuine American spirit too. And then the utmost worth of this novel will be lost. Some months ago I read *Presenting Lily Mars* by the same author, and there I was quite sure it could only be understood by a man with at least a little part of American feeling.

Most of German people think too complicated, everything must be logical, they think with their mind and not with their heart and blame nearly all feelings as "sentimental." Many already told me I was spoilt by the literature I read. Nevertheless I am glad they can't blame me as one being unable to think logical since I got the best notes (marks) in Philosophy and Psychology. They just cannot imagine that the roots of real mankind have to be looked for far deeper.

But I believe I start troubling you by meditations not being of the least interest to you. Let me just tell you yet that I took up a new pupil for a certain reason. The lesson I teach him will be called the "American relief lesson" or something like that. Naturally that hasn't anything to do with the contents I teach him. He probably never will know why I took him. The name is just meant for me and—you. The reason for it is the following. *[Here it seems Ms. Grover has honored Christoph's request that the name and circumstances of this new student be kept confidential.]* I was very happy that you enjoyed the little figures I sent you, and since I now found out a way to get you something not so easy available for Americans as cocoa or coffee, I will start to send you regularly packages with figures made and carved in the "Erzgebirge." The fee I get from this new lesson will enable me to finance this task and cover the expenses and will only be used for this reason. I hope I will give you some joy by it. I am very glad you get some joy caused by me.

Today we have Saturday. Already on Monday I will go and get some angels for you. They are not easy to get, but I am on friendly terms with the shopkeeper and he certainly will sell me some. Else I will go straight to the Erzgebirge. These angels are very, very lovely, and if you like them I will continue sending them until you have a big orchestra. There are over 30 little pals playing different instruments, from the harp to the trumpet. All kinds of musical tools are used by this heavenly choir being conducted also by a little sweet angel standing on a small platform.

We have to get the packages stamped at the customhouse; this is open to public visitors next Thursday or Tuesday, so I will post the first package latest on Friday. Please write, if the contents are according to your taste. I, anyway, will start to send regularly now. I am so glad to have found some mean to make you happy with.

Now there are a lot of questions to be answered. But I like to do it, since I am glad you care for my troubles and my family. I wished you could visit me sometime. My mother and I would make you very comfortable and I would show you around and we would go places together all the time.

James Hilton is an author I like to read, nevertheless I was able to lay hands only on 2 of his books: *Lost Horizon* (I wrote you about) and *Good-bye Forever*. Last one is a beautiful novel about a non-conformist minister, describing his dayly life with real art of poetry. Everyone having read this story, I guess, will take a more liking feeling towards his minister who isn't always, as many think (and therefore restrict) a loft saint but suffering by the same troubles and sorrows, only supported by a greater love he must possess in order to fill his task; also the teacher must have it. Generally both are very related-professions (kindred), both aiming to lead people to deeper roots of life.

In literature I am not very fond of the new style developed presently in U.S. and also in Germany (for example, Hemingway). They write too short sentences. Nevertheless I was impressed by a German short story I read yesterday (typical modern). I rather would like to translate it for you. Perhaps your circle of people learning to write poetry may be interested in the way we write. I will try to translate it very correctly and keep the short sentences in spite of my antipathy. If you care to use it, you naturally have to polish it up. I will include it in my next letter. But I am again telling something you didn't ask for. My thoughts are always dragging me to write about what comes into my mind.

About Konrad, my god-son? Well, I wrote a letter about him in this spring, it must have been about April. Maybe I addressed it to Miss Jamison. Konrad is three years of age now, but already hospitalized for half a year. He got infected by his father who died from tuberculosis. Best remedy, as the doctors say, is good food, since only the proper nourishment gives the body the strength to withstand and recover. Well, we all do what we can. But we haven't the money to buy food at black-market prices. About you mentioning to help him we got very, very glad. And because you ask I tell you he mostly needs fats, milk, sugar and chocolate or cocoa, latter ones aren't available at all. I wouldn't think it best to send it directly to the hospital. I visit him regularly and his family does it too. We could always take parts along with us, the nurses are very nice and will prepare them for his meals. Best you would send it to my address in Leipzig, so I don't have to carry it by train. And there I am already approaching the next question of yours.

Where I live? Well, I live on two places. Didn't I write you, I study at Leipzig (there is also the children's hospital). Leipzig-city is about 80 miles from Dresden and over weekends I go home by train sometimes. Students have to pay only one quarter of the normal fee so I can afford it. Besides—as you know—we have long vacations. I naturally spend them with my parents at home in Dresden. The winter-term will start at October 1st. My address in Leipzig (you may write me there too) is as follows: Chr. Haufe 10B Leipzig W31 Pistorisstr. 6 Germany Russian Zone. There I have my sweet little room, and do my own cooking, there I take the greatest part of the contents of your gifts too, because my mother says I need it there more, since I have to look after myself. Well, and she isn't wrong. Here in Dresden they eat a lot of vegetable my brother's future parents-in-law are growing. (There we are at the next question of yours).

My second brother is younger than I am, but he nevertheless has a fiancé already (I am a one-hundred per cent bachelor yet), the daughter of the gardener I worked for last year. I enclose a picture of this brother of mine and one of my parents going on a lorry through the same garden. My brother Ehrenfried first wanted to become a merchant for lumber, but after he met "her" he took fancy towards growing tomatoes and salad. He is learning to become a gardener now. "Gardener from love" (do you know the opera by Mozart?). Well, that was all. (My brother John, the youngest one is learning to become a clerk). I like to talk about my family, and if you want to know anything, just do ask.

I also was interested in your view of America and school-life there. I think the things you blamed (lack of discipline and no hard studying) aren't as bad as you think. We Germans had too much of discipline and you see where we are now. Besides you are a young country, the future lies in front of you. Young countries and young people have a lot in common, childlike actions, antipathy towards studying, lack of mental discipline, but also: growing strength and growing culture. Better to look forward into the future than to live in past times. There will be a time your teaching will be the same and even better than the continental teaching. How much I wish I could help you working for it.

Sincerely yours,

Christoph

P. S. Please, give my kindest regards to Miss Jamison. I wonder why she doesn't write any more.

In this letter Christoph opens up a bit more and reveals the toll the constant working, studying, and hunger are having on his spirit as well as the spirits of Germans in general. The widow who won 1,000 marks on the radio show would have had, if my research is correct, about $303, which would be worth about $3,850 today. For someone who had left everything in Poland behind, the reaction Christoph describes may well not have been staged. His reading of American fiction shows when he uses "ain't" and "dough." His benefactors will not approve!

August 24, 1949

Dear Miss Grover!

Today the postman delivered the package you listed on August 1. The contents are very precious and an effective mean of help to overcome the present weeks of specially high need and lack of food. It is nearly always like that at the time of the harvest. The stock of last year is gone, and the new supply isn't available yet. August is the worst month, not even potatoes were on sale for some weeks. We literally didn't know where to take the needed food from. So you may imagine how handy your gift came in and how big the help was you gave us. The joy rendered by your parcel was an immense one, and we were near to performing a big Indian victory-dance around the wonderful items spread out on the table. They were well-selected, they astonished and baffled us all the day, we couldn't think and talk about another thing.

By that I was reminded of a broadcast I listened last week to. The radio-reporter went to the home of a woman who had won some thousands marks in a radio-quiz. I think you know this kind of puzzles. People listening near their sets at home have to guess whether the sound broadcasted at a certain moment was caused by an organ or a mouth-organ or an orchestra-band; or who composed a certain piece of music just played, etc. The listeners write to the radio-station and there the right votes are thrown into a box and some are drawn the way you do with lots, and these are the winners. The first winner is always informed by a radio-reporter and this interview is broadcasted. The widowed woman when she was told to be the possessor of some thousand marks fainted and we heard cries like: Get a glass of water! Run for the doctor! etc. But everything was so natural I don't think it was staged or premeditated. After a while she recovered and was asked what she would start to do with this big amount of dough. Well, she [said], my children and I are refugatives [sic] from Silesia (Poland) and had had to leave everything behind, now I can grant them lodgings and outfits again. The quality and quantity of joy as your people in America get when becoming millionaires is by much cheaper means to get here about. I felt it when getting your parcel too. I think if all Americans knew they were able to create sunshine, happiness and thankful prayers in my stricken country, they would long to count themselves to those the shining eyes of utmost happy people are resting on. There won't be a moment in my life I will forget your love and assisting hand.

The bigger part of the parcel I laid aside in order to take it to Leipzig with me, it will enable me to start studying with high keenness and strength this fall.

The miracle of today was the coffee-dust. I think you Americans lead a wonderful life, everything is produced so handy for you; it would be a fairy-land to a bachelor like me, no wasting of time to prepare food and always having the best quality and also the wanted quantity to his disposal. Now I can imagine it's easy for your married couples to go to work both of them. There will still be left time enough to enjoy hobbies or to go touring the beautiful country of yours.

I was really struck by the enclosed booklet and guide of North Carolina. The pictures look like taken from some theater scenery. When I will have a profession of my own, I certainly will save sufficient dough to make a trip to all these places recommended there. I think these months will be the most beautiful ones of my life-time, since that's really a country as I like it, not the sky-scrapers-cities just telling of civilization but the vast landscapes looking like gigants [sic] to our small resemblances hereabout. Our "Erzgebirge" or "Sascon Switzerland" seem to be babies compared with your mountains. If I'd been that widow winning all that money I certainly would climb "Grandfather-Mountain" next year and enjoy the sight from that look-out. Some way to cross the ocean hence and back will be found, one could be hidden in a store-room or assist firing the engines. Wouldn't it be fun if I appeared at your meeting for people who learn to write poetry and read some poems of my own? Naturally they are written in German. I really started to write poems some years ago. They ain't so much but they came from my heart.

I also wrote some opera-texts for a gifted friend of mine endowed with skillness [sic] of composing music. When he will be older he certainly will be a success; at least we hoped it of both of us when enthusiastically being knights of the Muse. We went together to opera-performances and resolved to revolutionize the theater-public by our plays. Well, we were 14 years of age at that time and filled with child-like visions. He is now nearly 20 and a solo-repetitionist [sic] at the opera of Chemnitz. We don't correspond much since I dropped writing at the time my country broke down. I was changed a little bit at that time. I think many Germans were.

There is a big emptiness in our souls. It's a life more in dreams than in reality; but it's not the dreaming inspiring poets, but it's some kind of lethargic reverie. Naturally this feature is the inmost part of the soul and hidden by many functions, and only those strangers who listen with the strength of love will detect the roots within: not only in Germany, the whole of west-occident, the "Abendland" is some kind of dying at the part of its soul. And I don't want to.

Well, but I don't know why I told you that. What I wanted to tell was to assure you I liked the poems you enclosed, they fit (correspond) rather well with the pictures that raptured me. And I look forward to the edition you are going to send. I didn't think before, North Carolina was such a beautiful country since I fancied it a little bit desolate. (In spite of knowing the song: Dina, of the state of Carolina...). But now I am quite of the opposite opinion.

In my last letter I wrote about a fortnight ago, I promised you to translate you a short-story I read some time ago and was rather thrilled by it. I don't know whether it will touch you or not. It has something to do with the feeling of the inmost part of us I told you about. I will translate it later and I beg your pardon for deferring it to the next time, but I have to since it's getting late and I want to send off this letter today yet.

Well, I forgot to tell you I rather comply with your opinion that dry powdered milk goes farther. The skimmed milk is very fine. We haven't any, so it is of much use, there is no need of improving or taking whole-milk, skimmed-milk tastes rather fine. Powdered one doesn't get spoiled either, just the right thing for me using about three table-spoons each day only, I fancy.

It's a wonderful cooking or preparing cocoa by it. Many, many thanks! Sincerely yours, Christoph. Please, give my regards to Miss Jamison. Please tell me if you can read everything or if I shall improve the way of my writing letters.

Christoph continues to express his gratitude for Ms. Grover's packages.

Leipzig, November 20 1949

Dear Miss Grover!

Today I received a letter from home containing your letter of November 1 and the announcement that your precious and valuable parcel arrived safely. You can't begin to know how happy I was to receive both greetings from overseas. Your letter is a wonderful and very interesting one, and I will answer it day after tomorrow when I will have an afternoon off.

Just now I am in a big hurry, it is dinner-time, or lunch-time (as you might call it) and in 10 minutes I have to get ready to run for the streetcar, for I have to visit 2 seminaries this afternoon yet, and tonight there is an important speech delivered by a professor. But I can't wait until Wednesday to thank you emphatically for your kind thoughts and love and so I wanted to let you know at once how happy I am you didn't forget to remember me.

My parents put a list into their letter where I can see from what beautiful things you selected again. It's such a big help in overcoming our needs, I wished I could explain it with words. You are really a fairy-queen to me and my parents and brothers and to little Konrad who doesn't know how far all the beautiful things are traveled and shipped.

I enclose you the promised short story today, in my next letter I will say something about it yet.

In the meantime, I remain thankfully,

Sincerely yours,

Christoph

As I read this short story, it seemed familiar to me. I believe it was one I read in high school. It was written by Wolfgang Borchert (1921–1947) in 1947 and titled "Die Küchenuhr," or "The Kitchen Clock." Borchart died of hepatitis six months after the story was published. While the destruction Borchert describes is certainly applicable to Dresden, he is actually describing Hamburg, which was also firebombed with a loss of life estimated at 37,000, which was higher than the toll in Dresden. I'm reminded of Edgar Derby, who in Kurt Vonnegut's Slaughterhouse-Five was surprised to find an undamaged teapot in the ruins of Dresden and was executed for picking it up.

PARADISE

A Short Story

They saw him already in a far distance when approaching them, because he didn't look like all the others. His face looked rather old but from the way he walked one could see that he was only twenty years of age. He took his seat amidst them on the bench. And then he showed them what he was carrying in his hands.

"This was our kitchen-clock," he said, and looked at them one by one. "Yes, I found it, it was left."

He held a round kitchen-clock being white-colored like plates and touched the blue-printed figures softly with his fingers. "It hasn't any value," he said as if he had to apologize, "I know, and besides it isn't especially beautiful. No, it is only like a plate, so with, white, pure lac . . . But the blue figures look rather nice, I think. The fingers naturally are only made of sheet-metal. And now they don't run any more. No, the inside is ruined, I am sure of that. It

Won't ever go. But it looks as it always did."

He drew a cautious circle with his fingertip, he drew it round the plate-like clock and said in a low voice: "And it was left."

The ones on the bench enveloped by sunlight didn't look at him. One watched his shoes, and the woman stared at her pram. Then somebody said:

"You lost everything?"

"Yes, yes," he answered slight, "everything, just fancy, everything. Only this one, this clock was left."

And he lifted it again as if the others hadn't seen it yet.

"But it doesn't run any more," the woman said.

"No, no, it doesn't, it is ruined. I know that well. But else it is still the same as it ever was: white and blue." And again he showed them his clock. "And the most wonderful part of it," he continued excitedly, "I didn't tell you yet. The most wonderful part still comes. Fancy, it stopped at half-past-two, just at half-past-two."

"Then your home certainly was hit at half-past two," the man said, and moved his lower lip in an important way. "I heard it often. When the bomb hits, the clocks stop."

The other one looked at his clock and shook his head. "No,

sir. No. There you are wrong. *That* hasn't anything to do with the bombs. You mustn't talk about bombs always. No, half-past-two there happened quite another thing. You just don't know it. But that is the point that it stopped just at half-past-two and not a quarter-past-four or at seven o'clock. You see, half-past-two I always returned home from shift; at night, I mean. Nearly always half-past-two."

He glanced at the other ones, but these had taken their eyes away from him; he didn't find them. So he nodded towards his clock.

"Naturally I was hungry at that time, returning from work. And I always went into the kitchen and there it was nearly always half-past-two. And then . . . well, then my mother entered. How cautiously and easily I ever managed to open the door, she always heard me. And when I looked around for something to eat in this dark kitchen, suddenly the lights went on. Then she stood there clad in her woolen jacket and with her red shawl. And barefooted, always barefooted. And our kitchen was tiled. Her eyes were very small because the light was so bright. For she had slept already. You know, it was night."

"'So late again,' she said. More she never said; only 'so late again.' And then she heated my dinner and watched me while I ate. There she always rubbed her feet against each other, because the tiles were so cold. She never wore shoes at night. And she stayed with me until I was satisfied. And then, when I already switched off the light in my room, I heard her moving the plates yet. Every night it was that way. And mostly half-past-two. I thought it was quite natural and couldn't be another way. For she always did it. And she never said any more than 'so late again.' And these words she said every time. And I thought this would never cease to be. It was so natural to me, all that, since it had always been that way."

The time of a breath it was quite still at the bench. Then he said gently: "And now?" He looked at the others, but he didn't find their eyes. So he whispered into the white-blue and rounded face of the clock: "Now, well now I know this was the paradise, the real paradise."

It was quite still at the bench. Then the woman asked: "And your family?"

He smiled confused: "Oh, you mean my parents? Yes, they are also gone. Everything is gone. Everything, imagine, everything...gone."

He smiled embarrassed at them one by one, but they didn't look at him. So he lifted the clock again, and he chuckled dreamingly: "Only this clock was left. And the most wonderful thing is, it stopped just at half-past-two--just at half-past-two."

Then he didn't say anything any more. But his face looked very old. And the man who sat beside him looked at his shoes. But he didn't see his shoes, he always remembered the word "Paradise."

In this next letter Christoph is writing to a new benefactor who has evidently read his letters to Ms. Grover and does not approve of Christoph's use of colloquialisms. When Christoph refers to the U.S. as a country untouched by the miseries of the war, he is undoubtedly thinking of the fact this country escaped physical damage, but I suspect he may have been on thin ice considering America had lost more than four hundred thousand people in the war.

November 25, 1949

Dear Miss Frances,

Many, many thanks for your wonderful letter of November 1. It is always like bright sunbeams getting through the dark clouds having covered the sky day for day when some greetings from a country untouched from the great miseries come into a land where the war wrecked the happiness of men. Sometimes I rather would like to live in the rural country than in this city where three-quarters of all buildings are completely destructed, and I always am reminded of the terrible war anew. Only ruins are to be seen and a small account of houses being in order scattered around like autumn-leaves on the vast water-face of a pool. Sometimes I walk for an hour without seeing any house people live in, and some years ago some thousands were living in the ruins I am passing. So we haven't the real city-life as you might imagine. It's a rather gloomy atmosphere.

Can you imagine how it is to lose everything you possess by the bombs and to save only your naked life. Thousands of people did and live now the life of the past; they go through the present world as dreamers dreaming of the paradise they lost. They are retrospective, as you may put it, and all the strength they need to go living they don't draw from the future or the present but from the past. When you will fancy this feeling, you will understand the story I promised, you so long. I translated it some time ago already when I was rather touched by it. But as it often happens, the first impression went by, and I didn't think it so good any more, so I put it away, but since you ask for it, on Monday I enclosed it. I am no real good translator, either, I noticed it when I reread this story.

Please, tell me if people in your country remember yet their brothers and sisters suffering on this side of the ocean, or if they forgot our sorrows and stopped to pray for us. Very thankful I am for the kind deed of your friend. It shows me that the idea of bringing mankind together in love and understanding still is powerful enough to spread over this dark world and turn it perhaps in a bright one some day. Please, give her my best thanks.

The changes in Germany don't affect and hamper any post-delivery, I am so happy about that, too. I sent you a new box of angels I hope you enjoy. I am so happy you like those figurines and it's really no sacrifice to give some lessons in order to pay for them. I like tutoring; when I can give knowledge and self-assurance, education of cultural worths into the hearts and souls of children and their shining eyes tell me of their surprise when entering a sphere until unknown to them, I don't have the feeling of work but I would be unhappy if I couldn't do it. I really can't understand how you long to become a librarist [sic], certainly it's a very good and interesting profession and needs very clever spirits (I am always angry when I ask for a book and the librarist [sic] doesn't even know about its existence, less of its meaning), but it's really a higher art to awake the life slumbering in all those books and make it available to growing darlings.

It is a responsibility upon which depends the further development of state, nation and world. Certainly many don't appreciate it, many parents look down at the teachers of their children, but in that case I am sure those parents are to be looked down at. People should have learnt by now that teachers form the characters and the future of the coming generation. And when even the grown-ups know why they have such high-soaring ideas, such wonderful inventions, such knowledges of art and faith, they would see the roots of everything in their schooltime and today yet draw their hat in front of their old teacher who rests under the earth on some forgotten churchyard by row, perhaps.

My examination—in order to come to the point again—will be passed next fall. But I certainly will try to get the Doctor-degree afterwards. But as far as I now know, there will not be the money for this promotion available.

The box I sent you will get to you just at Christmas, as I hope. So I may call these angels the "Christmas angels" bringing you my best wishes and greetings personally. Our Lord may give you and your family very happy days and a blessed New Year 1950. Certainly, the Christmas-celebration of yours and ours will be a lot in difference. Our material base couldn't be compared with yours. We already are happy that we can buy candles this year the first time after the war, so we can put and adorn the usual Christmas-tree with its numerous burning candles. It will be very, very bright in our room, everything shall be light and shadowless, because Jesus wanted to bring light into this world of misery and shadows. Well, his light didn't conquer the whole world yet, but at Christmas we are supposed to remember that it depends also upon us whether we follow him and carry his light into the dark world, not only by lighting a Christmas-tree (which is rather a symbol) but by showing love to everyone. If there may be differences according to our celebrations and if you will eat turkey and other delicacies and we—well, I don't know yet what we will be able to get—in one point I guess we will have the same joy, that there was born man of God who brought light into this world and died for it.

In answer to your question, I'll tell you that the figurines are carved in the Erzgebirge—part of Seiffen (Annaberg).

A picture of yours I don't possess yet, but I would like to.

About the "ain't" I am very thankful to your advice. I thought it very fine since I found it so often in a Western novel. But I won't use it any more, besides I write a western-thriller, and there—as far as I understood you—I also can use "dough" for money. Naturally all these strange words seemed me to be rather alluring at first.

At last, there is something on my mind I'd like to tell

you. I hope it doesn't darken our friendship that I so often talk about our need and misery. It's always an uneasy feeling to talk about it, but I hope you will understand that different situations cause different ways of talking. I want to assure you that our friendship doesn't depend upon those trifle things which are only important as long as one needs them and lacks them but can't grant happiness. Certainly the soul can't rise without the proper treatment of the body, but by treating the body the soul never will rise automatically. The body is only the theater-stage, the boards have to be save [sic] and the scenery has to be correct, but all this doesn't make the performance. If the actors are bad, everybody will yell them down. It depends on how the actors act on the stage, and so it depends on how the soul acts in the body whether men can be happy or not. It's only the pity with us here that the boards are dangerous to play on, and this naturally infects the playing of the actor. But there will be a time they will be safe and sure again, but there will be no time I'll forget your helping me to carpenter them. And I would be happy and thankful if through all the years of our life coming yet, our correspondence would flourish and bloom, strengthening the spirit of friendship and human brotherhood.

Sincerely yours,

Christoph

P. S. The short story I sent some days ago. I hope you will have received it.

I wrote you about the happiness I experienced when getting your wonderful parcel, too. Again many; many thanks; it's such a big help you certainly can't fancy. I wished I could repay you some day. Perhaps I will be able later on; I would be happy to show you my gratefulness.

A new year and a new decade. Christoph's mood seems to be better.

February 9, 1950

Dear Miss Grover!

Having left the year of 1949 and going through a new one, our thoughts always are set on what we remember most and what we expect dearest. When the last night of the gone year enveloped me with its darkness, I had to thank my heavenly Lord for so many happy hours he rendered me and whose light also was spread over the dark ones, so I could see their real meaning which never was to torment me but always lead to the beginning of the experience of God's help and of love burning in the human hearts yet. When never been in need, one never will experience the true and real love of one's friends whose hearts are revealed in the situation when sacrifice is asked and no reward promised. Often I thought I had many friends and was happy about it, and then I happened to get into a jam and I noticed those I trusted most left me and those I thought to be not interested in me at all stood at my side suddenly. What blessing can be found in those hours when not only seeing the darkness but also noticing the small light flowing concealed from one human heart. It's a great rejoicing to realize that. And I am sorry for those who never entered dark hours, since they always will be of opinion to walk in bright light and don't notice it's all artificial; only a switch may be turned and lights go off. So I am thankful for the dark hours, too, for they gave me the friends I want to walk also when the light is shining again. When I meditate about the past, there's always you smiling at me with hearty eyes filled with love and understanding. You may-be never will know what strength your friendship rendered to me. So, when looking onward to the years coming yet, the year which certainly will bring improvement also to the countries most stricken by the war, there is one wish I most long to realize, to go on walking with you as good friends (in the real deep sense of meaning). And in spite of borders and oceans dividing us, we, will be nearer together than those who can touch each other but don't have any love.

My God-son Konrad is feeling quite content. A year already he is now a patient in the children's hospital and of the 4 children hospitalized at the same time he is the one feeling best. One of them died half a year ago, the other one is dying at present and the third, a boy of 4 years being the best friend of Konred, is making only slow progress towards improvement. Konrad is the only one being allowed to get up some hours each day. Soon he will have to leave "Dosen" and be transferred to "Klinga" where a branch of the Children's Hospital is situated, because the TB-ward is being dissolved. Only the ones who have no hope to recover stay here to die. But presently Konrad can't be transferred since in "Klinga" the measles are going around, and the children can't be brought into danger of infection. So he can be paid some visits yet, for soon it will be more difficult because we have to go by train. For him it will be better, since there will be verandas with deck-chairs and more trained children-nurses who are supposed to play with them. Here they start to blunder and do a lot of silly actions, they climb over the beds, race around in a big hurry, fall off their children-beds and hurt themselves. Recently he bumped his chin that way it had to be clamped, his friend broke his arm. Well, I don't wonder about it. It seems me quite natural, that those who are imprisoned into a bed for a year already but feel quite well are longing to undertake some bold enterprise. He asked me to thank the "fairy-queen" for her gifts and tell her about his living, so I wrote you these lines in his name.

Tomorrow he will cheer when I will deliver him your lovely "Teddy-bear" and some other things which certainly were meant for him. He is really a lovely guy, everybody *must* like him.

But I have already covered a great deal of paper without thanking you for your most precious packages, your very interesting letter and your nice picture I put on my desk at once. Now I will be reminded of you even more often. And it's good to remember kind-hearted people. Mrs. Blakely didn't exaggerate in her letter about you, I knew it all along, but I am glad to get it confirmed by another person. It didn't mean any difficulty to read her handwriting even if I sometimes had to stop and ponder a while; but that is no disadvantage. A letter needs comparatively such a long time to be written, and it is read in such a few minutes that one can't esteem it right when reading in a hurry. I always read your letters a few times until I know them nearly by heart, and I don't read them in the streetcar or between lessons, but at hours of meditation when I have the same rest and attitude as required while writing letters. How else could you understand other people?

I could never share the opinion of those who read the morning mail while breakfasting. One doesn't write letters neither in this condition; and it's like despising the time the sender sacrificed by writing when hurrying along the lines and throwing the letter aside afterwards. Through letters friends are talking to each other; if the other would be present personally I would offer him a chair, take a seat myself, lean back and look into his eyes thinking: "Talk, my friend, I am open to you." Why not do the same when the other talks by writing?

I am rejoicing you are glad to get those figurines. I am happy to have found a way to give you some joy. You really don't get the best of the bargain, for it's you who be of greater help to me. The worth is not what one gives but what one receives, and the worth is: with what kind of feelings one gives and receives. If you could imagine what happiness was created by your last two wonderful packages you certainly would say: Chris gets the best of the bargain.

This time I enclosed an angel you may hang on your lamp over the dinner table. I am just sorry not to have the proper containers ready so one of the last figurines broke, I hope this package will get through undamaged. When visiting a student whose parents receive packages from America I saw a pound-can of coffee. I thought it would make a good container and he promised to safe it for me when it will be empty. I will try it next time.

I surely would like to send such angels to all of your friends, but I haven't the needed money and it is not allowed to get foreign exchange. They only could send things being even more important to us than money (as for instance cocoa, tea or coffee) and I would gladly trade them for your friends.

I also will ask some teachers of mine to correspond with your father.

Many thanks for the book. I started to read, but since I got it only 2 days ago I didn't really start. I will do it on Sunday, when I will have rest for some hours. Well, so long.

Sincerely yours,

Chris

The Soviets detonated their first atomic bomb on August 29, 1949. Eight months later, Christoph, writing from Leipzig, notes that English is not being taught much at the university anymore. Most of the professors who taught it have left for the west. At this time, it was possible to travel to West Germany. As a result, he has changed his major to religion, noting that there is a shortage of ministers. Christoph also notes that Ms. Grover would not like what his fellow students were saying about America in their classes.

Phil Leask noted Christoph's request for "pineapple-fruits" and says how to get "Süd-früchte" was an obsession in the GDR right up to 1989.

April 15, 1950

Dear Miss Grover!

Many thanks for your letter of March 25 which came last week when spending the Easter-Holidays at home. Now I am back in Leipzig to start with my examinations, that will last just 4 months. I wrote about them to Mrs. Blakely, so you certainly will know the hard way I have to work during that time. I will be ready to become a minister this fall but may try to get a Doctor-degree by going on studying for some months afterwards yet; but that depends on my financial means and upon the urging of my church-bosses wanting me to start the sooner the better. There is a terrible need of ministers here because so many have left and not many are studying for it any more. But this need—though being the main reason—is not the only one I dropped the Anglistical studying by and by and reduced it to a minimum (not to nothing!). The other reason was the departure of nearly all professors teaching English here, who went to western countries or to Vienna. English isn't taught in the old extent at our schools anymore. A friend of mine wanting to absolve his final Anglistical examination last winter had to postpone it, since there was nobody to examine him. Also, there is a way to speak about America you wouldn't like when sitting in our classes as you thought to be so interesting.

Another reason for changing the idea of becoming a teacher is the fact that there is no "Religion" on the time-tables of our schools any more. We have no private or confessional schools, but only state-schools, so all schools are under the same order. The Religion lessons may be taught by ministers or helpers of the church, since the church is an organization separate from the state or any political institutions.

At present, I have to work very hard because all the students passing the examination with me were only studying Theology while I had to do a lot of other stuff besides. (So I passed the interpreter-test over a year ago, etc.). But I think I won't ever have to repent [probably repeat] what I learnt. People learning different things aren't restricted to a small point of view and acknowledge more readily the relative importance of many things, while otherwise one's mind is concentrated upon just one point thinking it the navel of the world. The best result one can get by all the learning is the growing knowledge that one knows nothing at all. That doesn't mean I want to compare myself with Socrates, but when having touched so many problems, one isn't ready and as fast as formerly to find always an easily adapted answer to all questions. We have a professor who at least once in each lesson drops the word: "I cannot decide here." When I began studying, I thought him—compared with the others who knew everything—to be, one who should rather learn than teach: today I know he is the wisest of all. Naturally that shan't lead to a skepticism.

Most people go the easiest way: either dogmatism (in *all* subjects, not only in religion. Dogmatism means to absolute one part) or skepticism. It is more difficult to fancy that the truth may be found between. It's the sight of all different parts in their *relative* importance under the unity not gripped at with our clumsy hands.

Well, but there are so many questions in your letter yet, I can't allow myself to go astray. I will try to concentrate myself and answer one by one. About correspondence with your students, I will ask around and send you the addresses you may distribute. Some of us certainly will be glad and thankful for this offer, but since I am just back from home I can give you only one today. It's a student of Theology living in the same boarding-home with me, he speaks some English, too. His name and address: Manfred Hausstein, 10B Leipzig, W31, Pistorisstr. 6, Germany, Russian Zone. Anyone wanting to correspond with him may drop him a note.

Konrad is now transferred and his best friend is still with him. We hope they will improve more and more, until they may be able to return home healthy and strong. Naturally it isn't just a question of food to make them well, since such a big TB is a sickness grabbing very deep into the bodily construction of children, but certainly good food helps to improve faster and often brought the sickness to a stop. So I and Konrad's mother are very thankful to you for your help. Same as you, I think it better to put the things meant for him in a separate box, though I don't need to mail it, since I visit him sometimes and his mother goes all Sundays and often also on Wednesday. We have to ride a train for an hour only, so it isn't "out of the world." But I guess you also will have to regard the postage which might be lower when you put all in one package (but mark it). Konrad's "best friend" knows you quite the same by now since they always share your delicious gifts. The nurse asked me and I consented since I couldn't have allowed to torment one little guy by the sight of seeing the other one getting something he longs for in vain.

But I am glad you are of the same opinion.

The box you sent at the first of February arrived save and sound quite some time ago already. I hadn't to pay any extra postage on it. I think you will have got the announcement of its arrival by now along with my Easter-greetings. I don't know how I ever will thank you and Mrs. Blakely for your big kindness towards me, you just overheap [sic] me with your riches. What are those small figures compared with those values you sacrifice for my sake. And it's not only the material help itself I have to thank you for, but mostly for the strong faith to human kindness and love you keep awake in me. I hope you will know what it may mean to be able to believe in human love. The world has a different face then, there is no absolute pessimistic blackness and there is the own longing to keep this light glowing and enlarge its realm. The world isn't ripe to burst apart yet. (1. Mose(Genesis) 18, 26).

Also for the poetry book I want to thank you again. I learn to like it more and more. Just now with the arrival of the spring and the awaking of the nature, many poems touched me deeply. I wished I was a poet to put all my thoughts in music-like words being more equal to the beauties of this world. Until 5 years ago I often wrote poems, they certainly were not very good, though a friend of mine being a gifted young composer wrote some music to different ones. But this seemed only the spell of a youth-phase fading off in the time of life. I guess I wrote you about it already.

At last, some words about the figurines. I know they are exported to different countries to be sold there but cannot say if America is among them. They are very hard to get here since they are exported, that is what I was told. But certainly there will be some way to get you some, though it seems too difficult to get them directly from the ones who do the carving in the Erzgebirge, since there is only a custom-office in Dresden and Leipzig and they would have to go some hours by train. But the other ways you suggest may be realized. I went to the shopkeeper and she agreed, only she didn't want to send and mail it herself, there are questionnaires, etc., troubling her; besides she doesn't know any English. So she gave me without payment some angel-figures to send you; anyone who likes them may take them.

She would like to get something which isn't to be had hereabout (coffee, cocoa, tea or pineapple-fruits). Please, let the things intended for her be put in a separate box or mark them distinctly so there will be no mix-up. I asked also about Christmas figures like Mary and Joseph, etc.; but she hadn't any of the same kind as the angels, there are only big ones and those being carved in another way. They may be had at Christmas-time.

Well, so long for today, I have to return to my books.

I remain with my best wishes to you and your family in thankfulness,

Always yours truly,

Chris

P. S. Being at an opera performance of Handel's "Julius Caesar" I met a fellow student, who studies Anglistic and also would like to correspond. His address: Hans-Gert Kupferschmidt, 10B Leipzig S3, Arthur-Hoffmann-Strasse 15F, Germany, Russian Zone.

In the next letter Christoph describes the effort his studies require as well as his becoming a good friend of Emil Fuchs (1874–1971), a pacifist Lutheran minister who would eventually join the Society of Friends. Becoming associated with Fuchs will be a great booster for Christoph's career. Christoph is coming to some conclusions that are outside mainstream Lutheran practices.

June 24, 1950

Dear Miss Frances Grover:

Your very welcome letter just arrived, and since I else might in the stress of work put off answering it for weeks until my examinations are over, I will lay aside the "Biblia Hebraica," I was reading in, and send you some words of thankfulness and friendship at once. It may not turn out to be such a long writing you are accustomed to, but you will get compensated next time.

Our examinations are just that way you couldn't fancy. They must be quite different from yours in America. But also here in Germany the Theological tests are the hardest ones, and already one of the group, passing the examination this summer, did give it up and returned home. A very detailed report about the way it is handled I gave in my letter to Mrs. Blakely, so—since I guess you exchange my letters—you will know the answer to all problems by how.

Nevertheless, I will give you a short summing-up.

As I said, our examinations last four months. At the first day we get a theme for the scientifical homework which has to be turned over after 6 weeks. I got the theme: "Jesus' thoughts about Grace and Merit according to Matthew 19:16 to 20:16 and parallels." I was so interested in this topic I wrote over 150 large sheets. Many books I read for this in the libraries of our university. The outcoming was a point of view rather different from the one the Lutherans—our seminary trends in that direction—are accustomed to take; but it became very clear to me, that the way to become a Christian doesn't only depend on God alone, but there is a big part of human activity in it. Jesus always speaks of the *doing*. So the opinion of my Lutheran friends (tho I belong to a Lutheran church, my dogmatical conception of faith often differs from theirs) who think it only depends on God, whether our church will live or die, I cannot share.

Therefore the special activity, claimed from preachers, must be to speak in a language which will be understood by modern people and not in phrases about dogmatical conceptions having been important in the 16th century. People shake their heads and come to the conclusion that the Marseistians [meaning unclear] are right in saying the church is an institution of old times no longer needed now. Same with our songs. There is a tendency in our church to reestablish the old songs of the 16th century and the musical antiphons of the Gregorian tunes, but to banish the songs of younger times.

Well, this music may give a lift to those who studied it, but not to common people, who will again get the impression Christians know only to live centuries back. But to this argument of mine they answer: It doesn't depend on us if people will come to Christ; He will force them, and if he doesn't—all our preaching and singing is in vain. Yea, it is sin to believe by doing so, we could do something for Him.

So no songs, who "give a lift" can be allowed, they render subjectivistical and sentimental feelings, but the songs must be adequate to the objective Bible-word. And that is the Gregorian tune.

That the people of that time enjoyed their songs in the same subjectivistic way as we do with ours, they won't understand; they just can't see, that art is developing and each century has a "cultural soul" of its own. Besides, how can anything sung or spoken by men be objectivistic? Even Jesus spoke in parables, since this was the way people of his time were used to explain and argue in.

Here I made a good friend with Prof. Emil Fuchs, who lent me books of Quaker and other American songs I at once learnt to like and certainly will practice in my church. I know

Prof. Fuchs very well; we often talk about America; and he still speaks about your country with the feelings of friendship and love (not about your government). I will greet him from you and remember your sister Hester, when I meet him next week. But—though I am on good terms with him—I must say that I cannot understand his friendly and familiar opinion about the Marseistian faith. But he may change it being here for a while.

Now back to the examinations!

After 6 weeks we got the themes for a sermon and a "Katechese" (religious lesson for a school-class). I had to preach about Roemer 8:19-21 and to teach about Matthew 6:10. This is over by now, too. Nevertheless there are no marks published yet; they only will be at the end. Now there come 11 oral and written tests with always some days intermittent. 2 tests about:

Old Testament (a translation and interpretation of a Hebrew-text, questions about historical, cultural, religious facts of old Palestine, Aegypt, Assur, Babel, etc.), then: New Testament History, Systematic Theology, Practical Theology. And one test in Philosophy. Then all will be finished.

There is a lot of work; but there is also a big helper, and that is you. There are no words I could shape my thankful feelings in. Since months already I don't need to drink any surrogate-coffee or herb-tea and I have good food too. It gives not only strength to the body, but also my spirits get strong and awake by it.

All your boxes arrived safely and sound. The two ones you mailed on May 13 I received a fortnight ago, when being at home to talk with my father over the topic of my sermon. There were so many splendid things in them; they mean riches to us, people of your country may not fancy of.

I hope you will have got the two boxes with angels, I posted in April. One of it is from the shop-keeper; you may do with it as you like. In June I posted 3 packages (two from me) to you and two to Mrs. Blakely. This time I selected some boys and girls carrying flowers. I would like to know if they please you.

Referring to the pictures I sent, you were right concerning the one of our dormitory (student home), but the other one was taken in a village near Dresden where I visited a friend of mine.

My mother plays the piano quite well. Also I do, but not perfectly at all; nevertheless hymns can be played correctly. Today I enclose some pictures taken one and two years ago.

But you may not have them yet, so I let make some prints of them, since I have no new ones.

My best greetings to your sister and Mrs. Blakely and all those friends, who know about me.

Hearty greetings from my parents, who sent your letter with a letter of their own telling me to give you their love.

Also Konrad and both of my brothers are greeting.

Besides dayly go over the ocean the thoughts of those who like your country. Good luck and God's helping hand for your work!

Sincerely yours,

Chris

In this next letter Christoph is elated that he has graduated summa cum laude and has been invited to study for his doctorate. He is also serving as an itinerant minister for several churches, which he reaches by bicycle. He gives us a taste of the underside of postwar East Germany—husbands deserting wives, moving to the west, finding new partners, and not supporting their families. One man moves his girlfriend in with his wife. Women had little choice in the matter. Men, as well as money and housing, were scarce.

The war had left the male-to-female ratio out of balance. Phil Leask writes that there were 7.3 million more women than men in Germany at this time and the greatest discrepancy was in the twenty to forty age group. One of the subjects of his book Friendship without Borders, *noting the situation, writes in 1947, "… it always really makes my blood boil when some representative of the stronger sex thinks that with a pair of nicely ironed trousers, a well-knotted tie and a fancy hairdo he has given me something to write home about."*

We also learn the exchange rate is 6 East German (GDR) marks to 1 West German (FRG) mark. The dollar was worth about 4 FDR marks.

Konrad continues to improve.

August 25, 1950

Dear Mrs. Blakely!

Dear Miss Grover!

Since I am using my father's typewriter and since both of you are interested in the same happenings around me, I made up my mind—in order to write not twice everything—to take two sheets instead of one. It might not be right, but please don't be angry about it.

There are good news to tell you today. Let me start with the most wonderful ones that reached me only a few hours ago.

Since during the past weeks and months my time was taken up with a heap of examination work, I didn't go on visiting Konrad or his mother regularly, but sent the cocoa and the food I got from you and last time also some children's clothes I got from a church collection packed away in a box via an acquaintance of mine to Konrad's mother, asking her to take it along on her next visit to the children's hospital. Thereupon I received a letter from her yesterday, informing me of the safe arrival and the quick transferring of the goods to the addressee meant. In her words of thanks, she underlined again and again the joy of the little lad and his fellow about the delicious gifts. Most of all he enjoys the cocoa-syrup, saying it tasted like nothing else, not even the other cocoa in spite of being packed by the same firm. For this I can't think of any other reason than the lack of vanilla being not available here, so we can't mix any into the cocoa. You see the little fellow has developed already a very distinguished taste. He says when he will be big, he will visit you in a vessel of his own (he never saw a ship, less an ocean), and drink with you a barrel of cocoa-syrup and eat pineapples instead of potatoes and bread. Well, he seems to fancy your country more a paradise than I do.

But the hope to visit you some day isn't any more or less now than the slightest chance since his mother wrote me he will be probably discharged from the children's hospital in September. His TB is about to get calcified. His return home will happen to be after an absence of about half the time of his life; one year and 8 months he was among strangers who by and by became the most related ones to him, and his home he craved for so much in the first time of his confinement may by now have no real imagination in his mind any more. He certainly will miss the many playmates he became accustomed to. But his mother plans to take some weeks' vacations in order to spend all the days on him and afterwards I may take him home for a week. We have a nice garden with some apples, plums and pears, so he will be able to run around in his steam-engine manner eating fruits and not be confined in a small apartment of a city-house.

Well, the main thing is, he will stay healthy and become not the victim of a relapse. We all are happy, glad and thankful to have him back among those who walk in the light of life and not in the shadow of the approaching death. As soon as he will be released, I will take some snaps of him and send you. You certainly will enjoy them, for he is an amusing fellow.

These are the good news No. 1. The next ones you already will have guessed since during the last months I didn't seem to talk about anything but my examinations. At last they are over by now, and I have passed them—as did not happen since the war—with *summa cum laude*. I was really surprised, for I didn't imagine to be successful in such an excellent way since I studied so many things besides Theology. But the more I ponder about that problem, the more I become convinced that all the studying of other subjects wasn't a hindrance to becoming deft in one, but just the main reason for it. From a higher and broader view, one picks up the problems quicker and looks at them from different sides, so the lack of many single points and dates in one's mind is easily hidden by talking about the big lines. I know if I wasn't as lucky as to have got always at the problems and never to have got fixed upon single points that had slipped my memory or hadn't entered it at all, I wouldn't have been as good.

Well, it's over now; and I really join with Wordsworth into the song: "Up, up, my friend and quit your books!" for after the last day of the term in the beginning of August (it was fore-poned), I had the feeling of having slipped quickly down a mental decline and groped with the spirit of doziness through a misty world. I packed my books away and couldn't look at them for a while; only some American novels I had catched [sic] sight of in a second-hand shop I couldn't quit, and so—lounged in an easy chair being put on the little meadow of our garden—I drowsed and intermittently read the wonderful novels you might know: The General's Lady, Claire Ambler (Tarkington) and Arundel (Kenneth Roberts). Also your Nature Lover's Knapsack was placed beside me and the voices of the birds were mingled with the reciting of some poems fitting to the sunny and hot summer-days we are passing. Thereby my love for your language flared up again, and I nearly shed some tears about the providence having brought me to a state getting worse and worse in it instead of—as I had hoped before I gave up its studying—acquiring it as a second mother-tongue. Among my acquaintances there are quite some who like to speak English but none of them thinks it beautiful like French or German, therefore unfit for poetry and stage. Sometimes I really think I must have a bad taste since my opinion is so single among them.

I had hoped to go for some days on vacation to Berlin in order to see some American shows there, but since I had to restrict my money-earnings I was in lack of money; for it is very expensive to go to the capital because the currency or the western part of the city is six times as high as ours. For one mark we have to pay six ones, and since the mark-prices for movies, streetcars, etc., are the same nevertheless, it's a fortune one has to pay when wanting to enter them.

Besides, the new work already called me, and so I had to forget about vacations. You asked me about my future plans and so I will give you a small account of my doings.

I could have become the minister of a parsonage and so would have entered a job I could have kept for life-time. It would have meant a lot of work, since—as I told you formerly—a place where 3 ministers belong is filled today by one. During the war my father had to care for over 25 thousand inhabitants. What a strain of work! But I was asked by a professor of mine to work for the degree of a doctor (we call it

Promotion), that means to work privately on a certain subject for about a year or two. For this reason, I also could realize a pet-wish of mine for since I couldn't be put into a large parsonage, I am now providing the smaller villages with services. As far as I remember, I told you in a former letter about the many country-churches that are without any preacher on account of the lack of ministers in this part of Germany. People have to go to villages nearby while their own church is forsaken; that is a great pity to many as you will understand. So I swing myself on my bicycle on Sundays early in the morning already and ride by the power of my feet (there is no motor available) through forests, acres and meadows; push it up the hills and enjoy going downwards when I have not to turn the wheels. After some hours I reach the village church I had announced my arrival at, preach there, take my bicycle and go for another village. So the Sunday is filled. The country people are very glad and thankful and so I experienced most beautiful hours. It only would be fine to find an organ-player to go with me.

At present for two reasons the work is rather strenuous. First the weather is very hot and often I arrive damp with sweat caused by the bicycling being no sport at all sometimes. Then I am already exhausted before I have started yet. But your Borden's coffee always gives me the necessary lift to go on. Secondly, I am pressed by work since this kind is something new to me, and it isn't easy after the years at university to make a good sermon understandable to all and being of interest to everyone. At the first time, when I looked at my watch already 50 minutes had passed, me talking on the first part yet. Though I noticed nobody sleeping, I very shockedly hurried to the end and was glad nobody had left before I came to the Amen. It's a mistake of mine very hard to be overcome that I never find the end, and though nobody was complaining about it, yet I know the housewives are awaited for at home to prepare the dinner. A service lasting over two hours with all music is too long, because most of all the country-people aren't accustomed to sit on one place for hours. My preparations need yet a long time (perhaps that is the reason for the long sermons, too), so I am quite busy again; it seems always to be like that. First, one thinks: When this or that stepping-stone will be reached or this or that task will be done, I will have time to rest. But then it will happen that another task will claim your service, and there is no thought of rest any more—besides perhaps the hope for it, which always will be prolonged and adjourned. Nevertheless I hope, when I will have some training more, I can spare enough time to start with my promotion.

At present, I stay with my parents in Dresden, but with the beginning of the next term I often will be at Leipzig and so I already kept my room there rented, since I wouldn't get another room as convenient and proper as this very one.

That seems to be all about me being of interest to you. I just could tell you yet that we have a "vacation-child" with us at present. It's something like a custom here that people who don't live in the midst of a city or town (you know: in Germany also the houses of smaller towns aren't spread around with gardens between, but built tightly together with narrow lanes and streets) take a poor child from there to spend some holidays with them. Our vacation-child is named "Angelika," and she really lives up to her name, she is like an angel. She is 8 years old, has dark eyes and dark hair and has a beautiful voice; without stage fright and fault she sings the second voice to many hymns and songs, so we were hundredfold rewarded for the little we spend on her.

Her father—because he didn't get any divorce—fled to Western Germany where he now has a neat income and ... a girl-friend. He doesn't send his wife and his children any money or package of food or clothing. So Angelika's mother earns all the living of the family by being a typist (a much better one than I am). To the house they live in there is no garden, not even a yard, barring a shaft to let light to the back-windows. Therefore we invited the girl to us, and since she took a liking towards me I often took some strolls with her. Mostly she enjoys the Elbe River because her home-town is situated far from anything like a river. First, she always asked me to go with her to the "sea" and it took me some time until I [caught] that she meant the Elbe. Some days before, I took her for a voyage on a little steamer, and all the passengers enjoyed her joy. We only went to Meissen and back, but I bet she won't forget this event for years. I took some pictures, and when developed I will send you the copies.

For the copies you, dear Miss Grover, enclosed in your letter I am very thankful. They give me a good impression more of you and your family. It is also the fashion with our girls to trim their hair short. Your father certainly doesn't look his age. It must be really nice in your log-cabin of North Carolina (isn't there a song: Carolina gave me Dinah?) quite away from all the turmoil, and certainly only to be reached by your car. Or do you think she could go there bicycling from a station? I wonder what America would be without cars having vanished by some reason as it happened in Germany by the war and its close. Here we can just go where trains will carry us; but since there are train-tracks spread about all over the country, we can go nearly near to any spot when taking the risk of a good walk. Cars are only owned by higher people. Please, give my love to your father and your aunts. I wished I could have dropped in some day.

Many thanks I also want to render for the 3 packages that arrived recently (one a fortnight ago, two the day before yesterday). The joy you created with them was great as always since the need of those things you packed into them is still the same.

The shopkeeper's breath was really taken away when looking at those riches, and she promised me at once to do all she can to provide some angels regularly in spite of the hardness they are to get, since many are exported (but none to USA as far as she knows). She is also a poor soul. Somewhat over 30 years of age, she still mourns about her husband who returned safely from the war, but left her and got a divorce. He also went to Western Germany and leads a comfortable life with a new girl, without paying the money he is supposed to. Her situation must even be stated worse, because by the air-attack on Dresden her home was destructed completely and she lost everything: beds, boards, clothes, dishes, etc. Now she has to care for herself and her mother. So she keeps a shop (she doesn't own it, but the owner doesn't live here and pays her a monthly salary.)

You see we meet with misery step by step; some caused by the war, some by the bad heart of human beings, and some by other reasons. How many children I find on my trips who live in bad need, their parents are dead, or they live in concubinage with other girls or men. I [maybe] told you of this soldier nearby who, returning from war, forced his wife to share their bed-chamber with his girl-friend-WAC. So children grow up unprovided for, without God and proper family education and without proper food and clothing. The crisis of marriages is on its highest spot; there is nearly no family happy and content, as often they may try to appear like it to the onlookers, though. Only the short time fulfilling the task of a minister, I looked deep into the eye of heart-renting misery. And I thought I knew all of it already. I always have to pray the Lord may strengthen me so I can keep on. But now it seems to be as with my sermons; there is no end, and still such a lot to say. Therefore, I will do as in such cases, and hurry to the end, answering your letter as quick as I can.

Certainly, on the study of population I had to take courses, too. It wasn't laid stress upon by our theological seminary, though, but as it mostly was with me to love those lessons only touching the outskirts of the claimed studying as being of special interest to me as, for instance, lessons about Mishna, Midrash and Talmud or about the science of nature, I liked extensively to listen to teachers of sociology, there are real interesting and our existence touching problems in it, and if you take a liking towards talking over one or another, I would readily accept this line.

Both the boys whose addresses I sent you haven't heard anything from your acquaintances. But I hope correspondence will be on the way now; I didn't see them for some weeks.

For the letter of Charlotte Dean, I was very thankful. She wanted to help sick children and our theological students. I am sure she will bring many hearts to a thankful glowing. I already told some very needy fellow-students of this generous offer being a new evidence of real Christian love over borders and oceans, regardless to nationality and race, and they were so enthusiastic I hardly could keep them from staying with me until the help arrived.

Well, all this will spread the truth that the way of peace and understanding between nations depends on their willingness to stick to Christ or to look for him in order to find him at last. So it is also your friendship putting strength into me and the renewed faith unto the heavenly father who makes all his children to real brothers and sisters. I wouldn't be the same if I hadn't met you.

Many greetings from my parents and brothers,

Sincerely yours,

Chris

Here we learn how appreciative the Germans in Christoph's circle are for the packages his American benefactors send. Coffee—even instant coffee—is precious. I was not aware Borden once offered coffee. Note that even containers are in short supply—Christoph asks for empty ones, although he notes "full ones would be better."

October 2, 1950

Dear Miss Grover:

I want anxiously to drop you a note at once since my heart is ringing with joy about the new greetings from you. Though no letter has arrived yet since quite some time, last week and yesterday by and by 4 boxes arrived from you. A hard longing pervades my soul to thank you without delay for this sign of friendship and Christian love. God bless you and your dear friend for this kind deed. The one package I transferred to the shop-keeper who will split it with the owner of the shop whom I talked to long ago already. So they will provide me the little angels again and send you also some on their own account. Only the problem of containers troubled me again. The acquaintance of mine who gave me the coffee-cans haven't got any at present; last time they already gave me some very old and a little rotten ones they used as containers for mushrooms a long time. Their last boxes contained no canned coffee, so I had to take the Swiftning cans you sent me. They certainly are very proper but a little bit too small, there is only space for 4 [pieces].

Therefore I resolved to ask you whether you could enclose some cans next time; you may put in empty ones. Naturally full ones will be better; coffee (besides tea) is the most rare thing; the best certainly is your Borden-coffee, so I wouldn't be happy if you shortened it; but you might shorten the amount of jam and flour-items in favour of coffee. Also the shopkeeper asked me to tell you: tea and coffee is the most important; you might send only these and nothing else. (Nevertheless she asked for pineapple too).

Perhaps you won't understand the longing of the Germans for those things most of all, at a time when other items, essential for life, are short too. But you might get an idea if I tell you that since the days of Hitler (Pre war time! *Please,* do read the novel *1984* by Orwell) we sip only surrogates which don't taste well and don't give any lift-up.

The coffee intended for me (also the tea) I split with my parents since they (mostly my father) need it too and are always enchanted when being told: "Today real coffee will be served."

Through the kindness of yours it is possible to tell them often thus and so my father gets inspired sufficiently to fulfill his difficult and strenuous task.

Our church bells start to ring. I have to keep an evening service, so I have to close in a hurry because I want to drop you this note (for it is no more than a note this time) at once.

Please, don't forget me but do like to remember me often!

Yours most sincerely

Chris

P. S. Soon I will write again and more detailed about my present life and doings.

In this letter we learn about Christoph's father's work as well as how some Protestant denominations are viewed in Germany. I found it interesting that Christoph's family is a member of a Lutheran Church that accepts members from other denominations and is not recognized by the (still) notoriously conservative Missouri Synod. Christoph's father seems to have come a long way since his days as a "German Christian." Christoph also comments on those who do not attend church but wish their children to be baptized, those who wish to be married in the church, and those who wish to have the church perform funerals. It's important to keep in mind religion was discouraged by the Soviet occupiers

Christoph writes he first met Konrad's family four years prior to this letter when he was a "young student yet just starting up," so he began his studies at Leipzig in 1946.

Christoph discusses his older brother's dealing with the government's production requirements for his farm.

Note also that Christoph mentions in passing the letter will be read by censors.

October 30, 1950

Dear Mrs. Blakely!

Dear Miss Grover!

Again two letters of [yours] have flown into my room and their greetings into my heart. I enjoyed them specially because I received them a few hours before I went to a concert where the symphony "From the new world" by Dvorak was played. Dvorak had composed it while staying on a visit in North-America, and as Tschaikowsky [sic] did when visiting Italy and using Italian Songs as themes for his Capriccio Italia Dvorak quotes some wonderful American tunes in this symphony he wrote in the new world (as we Europeans call America). So the theme of the slow moving second part consists of the melody "Going home." I never memorized all the words, but I remember an American picture very strongly I saw about 3 or 4 years ago. I believe its title was: "It started with Eve," and the deepest impression upon me was made by a girl singing "Going home." Only a few scattered words are still flowing through my mind, so for instance the sentence: "Mother stands waits for me, I am going home!" but the melody rings in my ear from the start to the end. For this reason, I like the Symphony "From the new world" (No. 5 e-minor) most of all and miss no occasion to hear it. This time all my thoughts—enveloped in the longing and yearning tunes—wandered to your home and I [paid] you a visit most vividly in my mind, hardly differing from reality. It is strange and not to explain how to become homesick for a country which isn't one's own home.

But I am already driving along a channel of thought again which—as did happen so often—will bring me to nothing else but reflecting on ideas pervading my psyche, and nothing will be said about the facts being of more importance to be talked over. In your letters you asked a lot of questions about my family and me and so I will try to give a review of it.

My father is only 53 years of age and therefore can't get the idea of stopping to work. Besides he doesn't want to because he loves his service to God and His people that much he couldn't think of staying lazy. And who should take his place when he left? It would be a vacant parsonage more as we have so many already. I am only sorry for him, since his work is such a strenuous one you hardly will be able to imagine. I often wonder how he manages to carry on. To his church-district belong 22,000 (twenty-two-thousands) people, an enormous number, even when you have to realize that many of them don't visit the services any more. They are members, nevertheless, and want to get baptized, wedded and buried. Normally each week in our church 20 children are baptized, 10 people are buried and 10 couples are wedded. Besides the church has to do all the religious teaching; furthermore the sick ones call for the communion, the troubled ones call for advice, help and strength; three times a week congregations are held for those who want to listen to a special sermon—and pray together. We have the baptism of children and not of faith, I am not quite certain to be the right way but it is done since centuries. Nevertheless it is a problem—worthy to be thought over—whether we should now go on baptizing babies whose parents don't visit the services any more but only want to be buried by a minister because they are members of the church yet, not having declared their retirement. Perhaps the children will be sent even to the religious lessons but since they live in the surroundings of a not-christian-thinking home and school they only come while being ordered and forget all about faith and church after their confirmation. So the example of some small denominations (Baptists etc.) who have only several hundred members in the DDR (= Deutsche Demokratische Republik = German Democratic Republic = the Russian Zone of Germany) are a constant admonition to us. They are called "Sects" which word has no bad meaning but only says the not belonging to the common church of the country which is more destined by the decisions of the former kings or sovereigns whose subordinates followed them and had to become katholic, lutheran or calvinistic according to the creed of their superior until the 18th century when the lutheran Saxon people declined to turn katholic with king August "the strong" who converted in order to be crowned as ruler of Poland too. So we have not the different denominations mixed among each other same as in your country where the pioneers and first settlers already belonged to different creeds, being the dissenters of European churches. In Germany a certain territory still marks a certain denomination on the whole. So in South-western Germany where the sovereigns of the 16th and 17th century remained katholic or reconverted (counterreformation) most Germans are katholic while the North-west and the East has a different tradition. But nowadays a big change took place. With the millions of people displaced during and after the war, with all the [refugees] from the former German territory also their denominations, to which they cling according to the tradition of their forefathers, mingle with the denominations of those they have settled down among. Nevertheless the 3 denominations of Germany are still rather equally numbered; the Katholics, Lutherans and the Calvinists; both the latter ones were united as Protestants in Prussia and her former provinces. In Saxony, the Lutheran church I belong to is very broad-minded and many members of other denominations are readily accepted and helped, so the ones who disapproved this broad-mindedness separated themselves and are called the Lutheran Free Church. Only she is acknowledged from the American Missoury Synode, while we are blamed to be traitors of the holy faith. Baptists, Methodists, Mormons, Christian Scientists, etc. are of very low number here, because these denominations were imported from England, the Netherlands or America and reached only those who were serious Christians (not only Christians by name) and also dissatisfied with the home-church.

In Leipzig the Baptists and Methodists have rooms where they assemble and keep their services. I often liked to visit them since I am always inspired by a congregation of people who live up to their faith in such an example-like way. There only those are baptized who really believe and want to render their life to the Lord. In our church millions are baptized who will fight the church during their life, blame and curse her, or at least be indifferent to her existence.

Well, I again left the topic I wanted to talk about. But I hope you will easier by this understand the hard way my father has to work and under what kind of circumstances. As far as I remember I told you already once that he owned a car formerly, in order to fulfill his task and go everywhere he was needed. Sut [sic] now he has to go all his ways by feet or street-car. Sometimes he uses my bicycle but he isn't that conditioned any more to ride it without leaving an anxious feeling and waiting within my heart. He is such a good dad and his sacrificing life for the service of the Lord and the poor ones makes him rather a holy example to me.

Naturally a *[I?]* give him a big share in the coffee and tea you sent and he often longed to thank you for them (he needs them even more than me) personally with his own handwriting, but he besides the words "yes," "no," and "alright" he doesn't know any English. At school he learned beside Hebrew, Greek and Latin some French which were of pretty use to him when in World War l he was stationed as a soldier near Verdun in France. But today he also forgot most of it on account of stress of the other work, as much as he tried to catch up with it again. He even bought a new dictionary once, but I doubt whether he found the time to look into it once.

Well, that's about dad, who bears beside those sorrows also the worrying about Ehrenfried.

Did I tell you he was married last year? His wife Hertha, the daughter of a gardener whose garden will be inherited by the daughter Magarethe, possess a garden her grandfather donated her directly when being a child yet. So it was leased. Last year they took it over, but since the lessor owned all the animals and all the tools and implements and took them away with him Ehrenfried is now hampered in his work seriously. The missing items can't be bought at all presently or are extremely expensive. For this reason he cannot meet the competition and has to strife with all his young strength to fulfill the obligations put upon his garden. For—as you know—farmers and gardeners can't just grow what they want and as much as they like but there are certain regulations they have to oblige. A destined amount of this or that item has to be grown and rendered after the harvest, so the food—planned by the government—will be at disposal. Four weeks ago Ehrenfried had a break-down and was transferred to a hospital, but after a fortnight he got released and is back at work now. The harvest is over at present and he pulled through, this year. My mother—in spite of her weakness—often went over (his garden lies just across the Elbe River) and helped him; also his wife is an extremely good and busy worker. They got up at 3:30 in the morning and went to bed near to midnight. During winter we hope they will recover sufficient strength to pull through next year easier.

Another sorrow we have to beat about John, though he is doing good progress at work. He is a clerk at a church-office and has to keep the books of a grave-yard in Drestlen-Friedrichstadt; besides he does jobs as there are in a church-office as for instance take care of the registers of baptism or weddings. He likes his work but is constantly in fear of a stroke or fit of some epileptical appearance (though it is not epilepsy but is caused by his quick growing-up). Once or twice a year he is unconscious for nearly a day. Perhaps this sickness also results from the terrible air-attacks he [lived through] in Dresden; he got the first stroke then. The doctors say it is some defects of the nerves. [Today John would probably be diagnosed as having PTSD.] So my parents have to carry a hard burden, and I often feel ashamed for being not thankful enough to be best off of my family. I am perfectly healthy and happy at my work in spite of the tragedies I encounter often. But it isn't the happiness growing from a strength in my heart enabling it to meet need and distress without absolute despair, it is the certain and blissful feeling that one can't stumble and fall deeper than always in the hands of God who stands behind everything in spite of the dark veil often wrapped around him. Nevertheless a real shock happened to me last week and I felt at a loss with all my help I was supposed to render. A child—who was born after 5 months and nursed within a machine until being able to be given to his mother—became blind thereafter because his mother forgot to feed him vitamin A. There is no medical help; the eyeballs will run out within the next time and the child will grow up with empty eyeholes and never see the light of the sun anymore; the mother is despairing on account of her carelessness causing such mischief, always seeing the dark future of her darling. Why, why, why? That is the question shouted at me and echoing in my own heart unceasingly.

But there are also beams of sunlight. Konrad got released, and though he has to be taken special care of he looks very healthy and is a happy and content fellow, the joy to all who meet him. Contrary to my fears he feels very well at home and tries to spring up to my neck each time I visit him and bring with me some gifts of you. About his friend I will ask next time I go there. Konrad lives in Leipzig, about 90 miles from Dresden, there it is that I met his family four years ago, me having been a young student yet just starting up. It was an important story of my life and if you are interested in it I may tell you. His father died in the meantime so his mother had to work as a clerk but had to stop to do it since she has to take care of Konrad. She now lives only on charity and on her oldest daughter who is married to a medical Doctor. [This daughter was Christoph's former fiancée.] Besides Konrad has a second sister about 12 years of age.

Pictures I couldn't take of him yet, since soon after his discharge from the hospital the weather turned misty and cold. The winter arrived very early this year, and all the days I was in Leipzig I couldn't take him out because we have to be careful with him and his mother doesn't take any risks. And an indoor-snap I couldn't dare to take because the little fellow can't keep still for more than the smallest part of a second.

But the promised pictures of Angelika I enclose. We had already our first snow, and the cold starts to pervade the walls of our homes. There is no refuge to Italy as you might go to Florida for the winter-season, even if I had the needed money it would be impossible. My room in Leipzig is especially cold (in summer it is especially hot) on account of being directly under the roof. For the shortage of coals we can't always sit in the warmth of a heated room. But it is already far better this year than during the passed ones. During the last winters I often worked in bed wrapped in warm blankets like the poor poet in the picture by Spitzweg. This winter I won't have the time to work in such a state; the driving of a bicycle will be of no joy while the weather is freezing.

Most my time I am living at home now, where we heat one room—the "tunnel-chamber" as I certainly reported you formerly.

My district I have to preach isn't situated towards Pirna but just the opposite direction (towards Meissen), nevertheless I know Pirna well, and am glad that there is a location we both know so well.

The boxes referred to all safely arrived. (Only the CARE-package didn't yet.) I split them as intended and prepared a lot of joy. I only wished you would get a small part of this joy from the little figures.

I don't know the word "Hummel" figures, so, dear Miss Grover, do tell me please defining the word what they are like. I would love to send some, but not everything coming from Germany to your country is available to me, since Germany is divided into 2 countries and it is more difficult to go from one to the other than to cross the USA-Canadian border. So anything produced in western Germany may not to be got in this part.

This time I sent you the first of a set of three big angels carrying candles. I hope you like them; they are especially valuable, but since I am no student any more—though I am not a rich man yet and still have to calculate very carefully about my expenses on living—I can afford to send them. The postcards you liked I ordered and will send them soon.

Well, I run out of paper, I guess. Since I used carbons again (because it wasn't blamed by you) quite a heap of paper I thought to go for weeks has vanished. And the borrowed typewriter is shrieking with anger furiously because it isn't accustomed to be treated by a fellow being so incompetent and clumsy to use it. Though I learnt to write according to the 10-finger- system, I never came to the point where I could dare to take my eyes away from the tabulature [sic]. I certainly would make a bad pupil to Miss Grover's teaching.

But now I guess I have to quit, the censor might be worn out by now. I only hope you liked this chat and so I say Goodbye for today.

My heart comes over to you, my dear friends, enclosed in all the words of this letter. Please, keep me in your dear remembrance and get the feeling to have a true and loving friend across the ocean, far, far away.

Sincerely yours always

Chris

Here Christoph describes the difficulties the family encountered when a CARE package was sent them. The package was not allowed in the Soviet-occupied zone, so it had to be picked up in (presumably) West Berlin, 120 miles from Dresden. Christoph doesn't mention any inspections the package had to go through, so it appears this is just a way to make it more difficult for packages to make it into East Germany.

Christoph's entire family seems to be suffering from bad health, which is understandable, since food is still being rationed and the city is still demolishing what was left after the war. In addition, coal is both polluting and scarce, and the coal available is lignite, or brown, coal. Only one room in the Haufe house is warmed. We can imagine the air was filled with coal soot, lead paint, asbestos, and who knows what else. Christoph writes that the family doctor told Christoph the coffee he is receiving from America is helping his father.

Note that on Christmas day 1950 Christoph conducted a funeral for a child killed by a streetcar. That must have been devastating for all involved.

January 10, 1951

Dear Miss Grover:

Dear Mrs. Blakely:

About 3 weeks ago I started to write a letter telling you about the safe arrival of the care-parcel which had to be [fetched] in Berlin since it wasn't allowed to enter our zone, but all the obligations and preparations due for a preacher at Christmastime didn't let it grow very far; I can't write a letter when only minutes are left to be used for it and my mind was troubled with all the difficult problems put upon me by the dozens of services and speeches I was claimed to. And afterwards a serious sickness took me away from my work anyway.

But let me start from the beginning. In December the Care-Agency wrote a letter informing me that since only boxes directly mailed and packed by private persons were allowed through our mail we had to fetch the parcel in Berlin. Therefore because I couldn't take a day off my mother went to Berlin to get it. It was a heavy and precious parcel enabling us to bake and cook nice things for the Christmas-days. All of us want to thank you for this gift. Many many thanks also for the other package which safely arrived three days ago. You effectively help us in a real Christian and altruistic way to pull through this time of need. Your help has become rather a part of our life which we couldn't imagine without it. Nevertheless we didn't come to the feeling that it was quite natural, but each parcel reminds us with the same eagerness as the very first one that the easing of our life is paid with a big sacrifice of yours. And that we certainly won't ever forget.

Long before Christmas I already got aware that just these topics I thought to be easiest formerly, were the most hard to preach about. All the ideas seemed to be commonplace. And over all: Glory of God and peace on earth! at a time when of both just the contrary was going on in this world. And then this big number of Christmas-celebrations and those many people who look at them as some means to render high spirits the same way as whisky does—means to lift them over the bleak reality for some hours. They sing the songs and hymns in order to reach floating feelings and wonder when they awake with a headache or get disgusted with those celebrations. And we should be able to celebrate Christmas constantly through all our life! For without the nativity of Christ the darkness of this world would be complete and omnipotent. But it's a mistake trying to get the right Christmas-feeling from the way of celebrating instead of leaving the celebration to be the way in which a feeling—rooted far deeper—will find its expression and shape.

At first I was so much afraid of all those many services and assemblies I had to speak at. But when the time approached, I got very happy and always found the words to give at least some gleam of the joy of Christmas to the burdened people. Only once I nearly had to despair. It was at the 25th of December at noon when in a wet and cold weather

I had to go to a graveyard because a child driven over by a streetcar had to be buried. There I had to talk to parents who had already bought toys and a Christmas-tree and Christmas-candles. And not only there but at many other places I had to notice that Christmas was only the excuse to weep and cry deeper and louder. Why so many tears and so much longing and so heartrending suffering! And then I had to preach at the last day of the year about the word: Until here the Lord has helped (1. Sam. 7, 12). Never I had pondered that long before whether to change a prescribed text. Is all the suffering only a way of education or testing of the Lord as most of our Theologists say? So many people are broken down and driven away from God by it. And who shall believe in the "Alloving" and—"Allmighty"?

But it wasn't God who lead Jesus upon the top of the mountain and showed him all the riches of the world and it also wasn't God who stroke the sick man of Gadara (Lk 8, 27) with that terrible illness. There are other powers ruling in this world and the last victory of God belongs to the future. But that is the help, that we know His light will drive away darkness at last, and heal the suffering and sickness, the lonelyness and longing, the despair and the evil. And we are called to fight with him, in spite of our weakness, and when we fall and get beaten, the last victory will be His, and that keeps us. And if we fall we always will fall into His hands.

It was at the New Years Day when I fell sick. I must have [caught] it at this burial in the cold and wet weather. Before I didn't imagine it to be serious, but at last it was. When I consulted our Doctor he told me something at that occasion I didn't know until then. He said I could be very thankful to have American friends for he had put the alternative upon my father, on account of a serious heartsickness either to stop his strenuous work or to find someone who could regularly send him coffee as keeping medicine. "To find someone . . ." he said because this item cannot be bought here at all. My father hadn't told me though? It was over a year ago. He only said once—as I wrote you—the coffee did him a lot of good—and I thought it had something to do with growing tired. Ne never asks anything for himself, he only goes around binding the wounds of others, like a medical man going over a field after battle. But you shall know that helping him means helping many, many others.

After I asked my mother to mail the two packages I had made ready for you I went to Leipzig at the 2nd January because there I have a room which can be heated.

Yesterday I experienced a big joy when a letter from home informed me that another parcel had arrived and besides a letter of Mrs. Blakely was enclosed.

Today the temperature has gone down a little, so I got up, in spite of not being allowed yet, but I wanted to write you this letter. I am a little bit weak. Perhaps you will have noticed it all along this letter. Else I am feeling quite well. The others—also Konrad—are going along fine too.

Most sincerely yours

Christoph

P. S. Just now a letter informed me, that another parcel has arrived. Many thanks! (I guess it is for the shopkeeper). I am feeling much better today and hope to return next week. The shopkeeper didn't send a package this time along with the ones I sent; she seems worried to get the right number of boxes. But I promised her for sure.

A still-ailing Christoph is grateful for the Christmas packages from America.

January 24, 1951

Dear Miss Grover:

I am writing this note in bed. Since a relapse caught me I couldn't return home but my mother had to come to nurse me. Well, I thought myself to be too strong and have too much energy.

But it's not serious enough to worry about. A great joy for me was your letter she brought along. I am so glad you liked the figures I sent.

From April as far as I remember I always mailed three packages, two packed by myself, one for you and one for Mrs. Blakely, and a third one packed by the shop-keeper. The figures *I* sent were selected that way they would be new to you, the ones *she* delivered me were of her own choice and often some you had already, but we thought it wouldn't matter since you gave those away anyway as you wrote. One of the set of the kitchen-angels (heavenly bakery) was from her too. Since you were so very fond of them I will try to get another set as soon as I am up.

Mother brought also a letter along I had written to Miss Dean but was returned with the remark "not listed." I enclose it and would like to ask you to forward it to her. She wanted to help our sick children and students, but I didn't hear anything of her any more. Now I know the reason was that my letter didn't reach her.

The coffee-jars always arrived safe, they must been made of very strong glass and are very well packed by you. They never broke but one, it was perhaps 2 months ago that one was broken, but the coffee could all be saved and used.

My mother told me how fine and wonderful the two boxes were that had arrived. One of them I asked her to take to the shopkeeper, so she certainly will give another package to my mother.

Some days ago a Christmas-card from Miss Jamison arrived. I was very glad to hear from her, that nothing had happened to her.

Now I have to close. My mother tries to wrench the pencil away from me since I am ordered complete rest.

Well, I will write more soon.

Sincerely yours always

Christoph

The following is a letter from Christoph's father, who expresses gratitude for the packages from America as well. He also describes his work as a minister in five Dresden suburbs as well as seven villages. He has 20,000 parishioners, most of whom do not attend church. The letter was translated by Christoph.

Translation

March 21, 1951

Very favoured and dear Miss Grover!

Since our Christoph abundantly was provided for with most valuable gifts of support through your love and kind mediation we both, my wife and me, are longing to thank you and your dear friend from all our heart through a personal note today. For everything Christoph gets he splits up with parents and brothers, truely and readily, so each box coming from far across the sea is received with great and enthusiastic rejoicing. For some time your cans of meat and fats were the only saving means to provide the badly needed number of calories to keep body and soul from collapse. Also now we wouldn't get along that well without your help adding to our rations. But as an especially precious remedy your instant coffee is judged upon. When soul and mind are dreading to break down or when there has to be made a point of some specially requiring task in my service and obligations a cup of real coffee is prepared at once and I am helped to pass the obstacles arisen, also to get rid of the first symptoms of a growing calcification of the arteries and the troubling headaches caused by it. So it happens your gift of coffee is also of profit to my church-people. But in quite another way too! How often people call upon me who have become absolutely poor and never imagined it possible to drink a cup of coffee once again in their life. And now this wonderful delicacy is served to them. You ought to see their astounded reaction and their happily glowing eyes. In Germany we had to live through years during which we worried whether we could satisfy the nagging hunger, weather we could manage to get sufficient black flour, bread and potatoes. And now we can sit at a table served even with things like coffee, thanks to you! About that we certainly have to be happy and thankful.

Also the arrival of the box with the nice-looking bed-jackets seemed to be like a miracle, since we got them just at a time when they were needed most. My wife was laying down with the sickness of inflammation of both ears; she had great pains days and nights, and we couldn't heat the bed-chamber. But covered with your woolen jackets she could dare to sit up once a while. Wool is one of the greatest scarcity, there is no gram to be bought on tickets in spite of being essentially needed. My wife has recovered by now and wants to type this writing because she is afraid of you not being able to read the German letters. A carbon will be rendered to Christoph, so he may give you an English translation.

Your letter of March 11 arrived today. We wait for Christoph to return home the end of this week. At the beginning of next week, that is just after the Easter holidays, we will mail some packages to you. In the meantime I have a lot of work to do. Tomorrow I have to keep two services of communion, the day thereafter, the Good-Friday, I have to preach at a country-church which is without any minister, so I have to attend to her. At night there are communions again. Beside there are services as for instance golden weddings, etc. At Easter-Sunday I start at six in the morning with my first service, there also is a service for children at eleven and in the afternoon 10 children have to be baptized. Last Monday at a congregation of male people I delivered an address about the topic "the doctrine of evolution and the miracle of creation," and last Sunday I had to keep confirmation for 130 children. In Germany we still stick to the "Volkskirche," that is the great masses of people still belong to the church according to tradition, but many are only Christian by their names and nearly never enter a church beside the occasion of some family-ceremonies. How many people I have to bury without ever having seen them through all my life. My district is very large. Over 20,000 people belong to it, and it consists of 5 suburbs of Dresden and 7 villages now. Therefore I often have to walk far ways, and above all, since some month I had to take over the tasks of the deceased Superintendent of Dresden, too. I am writing you all this just to give you and idea of some part of my living and working. I love to be a preacher, for only by the gospel all our life if given the last aim and end. Without gospel there is no hope at all. Both of us, my wife and me, therefore are glad algo Christoph had decided himself for the profession in church and Theology. Concerning science he is more interested than me, also concerning language, on account of his gifts and purposefulness. Well, you will know him from his letters.

Therefor I close this letter of mine by giving you hearty greetings and the best of wishes, and remain including my family,

Sincerely yours,

Rev. Ehrenfried Haufe

> *Christoph is feeling better, but is still suffering the aftereffects of his illness, including brain fog. He's revisited* Gone with the Wind, *this time in English, and compares the South after the Civil War to conditions in Germany. A historian I know who is an expert on the Civil War tells me many people in war-damaged nations found common cause with the defeated American South as portrayed in that novel. Christoph mentions that twenty-five cigarettes are allowed in aid packages and asks for them even though he does not smoke. I would imagine cigarettes could be used for barter. I was advised to take Kent cigarettes with me when I went to Romania shortly after their revolution. That turned out to be very good advice.*

March 29, 1951

Dear Miss Blakely

Dear Miss Grover

Many, many thanks for your concern with me. It does a lot of good to know somebody is remembering me also in the dark days I had to pass, and the feeling of lonliness and despair is warded off that way. The one thing that matters is anyway for the heart knowing it doesn't beat for nothing but is of some meaning to the fellowman. So the most troubling thoughts during my sickness were concerning the impotence to serve and to leave so many people waiting for me alone and perhaps letting so many slowly growing seeds I had thrown out under hard labor go to wilt and rot because the needed nursing for further growth was lacking. There aren't any ministers left to take my place and though my father in spite of his constantly accumulating obligations tried to tend also to some of my forsaken district he couldn't, naturally, be everywhere, though working day and night, and he ought to have some rest too. I hope to give him some holidays when I am recovered completely and able to overtake some of his tasks, for he isn't one of the strongest any more but should take care of his health and stick more to some work in a heated room than to run about in the countryside while snow and sleet is smitten into his face, But he is such a strong believer and his faith puts such a lot of energy into his weak body he is the very well of strength to me too.

It was a rather bad flu I had [caught] this winter, for when in February I tried to return to work all of a sudden I was blown over again and had even to be hospitalized for some time. The second relapse just happened at the day I mailed the "Rabbit-School" to you. I was overglad that morning about everything going OK with me, and with all this joy I had to send you something and wanted to write you a letter the other day and go home the next. But at night the temperature had returned and everything was as bad as before. Well, it wasn't too bad; I hadn't any pains and I never ceased to be thankful for that. I had visited too many people writhing with pains and calling for death to free them I didn't forget the blessings to be found in a sickness without any pain; and though I grew rather shallow-looking and unable to direct my thoughts the way I wanted, I can't blame for an unpleasant feeling to be drifted away in some feverish fantasies. And when recovering I could tend to many things I had to neglect the months before. I could read some rather fine novels by Booth Tarkington (*Women* and *The Gentleman from Indiana*), Hemingway (*For Whom the Bell Tolls*) and—because I was forbidden to work on scientifical books—started even to read an English edition of Mitchell's *Gone with the Wind*. (I guess I told you already that I knew this novel from a German translation I was ravished by at a time shortly before the war having been a school-boy yet. Then it left the impression of a far-away world like any adventurous novel and only when the war had started and came to such a terrible end I knew the drawn happenings to be the reflexions [sic] of reality. But never I grew more certain Miss Mitchell must have been a great author seeing the things of life than when me enliving [sic] the fate of Scarlett O'Hara at a time when all the told miseries had already touched my life. The blockade, the bottling-up of a country, so no necessities of life could be shipped in, the shoes with soles of wood, the hunger and the prices going high to fantastic figures, the graybeards and the little boys marching to war and death, the bombing of the sieged cities, the typhoid and the way home walking along ruins with blackened chimneys and fancying the house of childhood being a forsaken heap of rubble too, the anxious hurting to the bend of the lane where the home could be seen; I had to go through all that, and million of my country-men too—what contrast to the happy life before, the all-day-parties at Twelve Oaks and the crazy cheers and illusion at the beginning of the war. Shade becoming more impressive when seen compared with light. Scarlett returning home from Atlanta to Tara, I read these events just last night and the similarity is still haunting me. And when the soothing feeling of remembering that everything belongs to the past starts to take away and dim the huge meaning of such miseries the news from Korea rent my heart anew.

I am sorry Miss Mitchell had to meet a tragic death too, she knew the features of human hearts and all the suffering mankind is able to bear; what difference to Kathleen Winsor and her "Forever Amber," latter one. might know herself rather well (and it's a great thing to know this part of reality and the eyes of many women might have been opened by that book) but she fancies everybody like herself.

At the Easter-Holidays I returned home after I was severely admonished to stop the past kind of work, all the bicycling making hot and sweating and being of danger when afterwards speaking in cold rooms and churches. I am not of iron but of flesh and blood and have to live according to it, etc. Well, what should I do? Doctors have mostly right about the way to treat a body when wanting to keep it strong but are they also competent to decide whether it isn't necessary to risk something for a higher value than just taking care of oneself? At least Easter I couldn't stay home but had to go around to those who waited for me such a long time, and there was joy and thankfulness everywhere. I never thought I could have won so many friends in such a short time. I guess I can't leave them. I have to remember the preachers of the West at the American pioneering-time when the first settlers were pushing forward in their carts and wagons, cut the vast forests, plowed the first fields and fought the Indians. Then there also were some ministers who drove from settlement to settlement where for years so church-service had been held and baptized the children, wedded the couples, taught the right way of life, and went on their route. Well, it isn't exactly to be compared with my task having to be done under quite different circumstances and conditions but I was struck by something of similarity between them and now.

So I started to go on my way too. I am only wondering about me being somewhat different from formerly. There seems not to be the same energy; I don't mean the bodily constitution but the mental ability to rise and spend itself with a creative power all day. Often when starting to work for some address I don't come far beyond the first sentence and when I hard try to go on all the words seem me to be bleak and without living, constructed by calculations; and so I find myself pacing the floor, impotent, or looking dreamingly in front of me nowhere. But I hope everything will rearrange itself. My mind may be just like a man who had a broken leg and starts his first paces after the long process of healing, he has to walk slowly first and rest at intervals.

When I returned home a big joy was prepared by your letters placed upon my desk by my mother who just had recovered from some inflammation of her ears. I read them over and over and was so happy you didn't forget me, though I couldn't write for quite a while. Besides the one package of the two mailed in February had arrived already and so we enjoyed our family-reunion at a splendid Easter-Coffee and Easter-Tea. I at once went to the shopkeeper and asked for the items mentioned in Miss Grover's letter; but I at first could only get the "Pied Piper of Hamlin" and also some beetles providing the ones seen at the picture though. Letter hard shelled bugs and the set of Mary and Joseph I ordered at least and as soon as they arrive I will mail them. The trouble with all those things is their scarcity, for instance those little angels are nearly nowhere to get, and if they are, customers get only 2 ones at a time. So I am glad I get them through this acquaintance of mine and I am also glad she sends you some on her own account. For I am longing to give you some joy, and since I see you loving those little darlings I am glad to be able to send you some. Don't be troubled about the money. Certainly I don't earn riches. Presently I get 130 (One-hundred-thirty) Marks each month. Naturally that may not mean anything to you if I wouldn't give you some example of the prices to be [paid]. All the rationed items are not very high. So I pay for 50 grams of meat about 30 Pfennigs (100 Pfennigs = 1 Mark) and for a round of sugar (we get 3 pounds each month on tickets) 56 Pfennigs. But all not-rationed things are very expensive. A typewriter costs 750 Marks, a quart-bottle of whisky 30 Marks, a pound of sugar without tickets 6 Marks, etc. Well I don't need buy any of the latter ones and so I get along in spite of the additional pay for room and light. Since I haven't the spare time to go to many entertainments and besides don't smoke I save a lot of money I don't have to earn. Naturally I would like to buy a good suit of clothing once, but on account of not being able to spare the needed 350 Marks I am not sorry to go on without. But I would be sorry if I couldn't send you those little figurines you got such loving feeling towards, and so always sufficient means are left to go on sending them. I only have to pay for one package each month, anyway. Though I always mail 3 ones, the first is paid by my acquaintance, the second by my father and just the third by myself. I also split the things you send me with my parents and they are that happy about your gifts my father wrote you a letter of his own to show you his joy and thankfulness for this efficient help of yours. And my mother was very glad about the woolen jackets and Konrad about the cocoa and the sweater they just yearn to thank you. The things for Konrad I take along to Leipzig regularly. I have to go there rather often (it's only 70 miles from Dresden) since I only there have a heatable room. In Dresden just one room can be heated because only one stove could be built into the flat which formerly was installed with a radiator (heating with hot water running through pipes) which cannot be put to work at present on account of the shortage of coals. So I can't get the needed concentration for my preparation-work in that room crowded by the whole family and visitors constantly pouring in. Naturally during the warm part of the year this difficulty does not arise, but also then I have to go to Leipzig, because I often have to look things up in books only to get at the library of Leipzig, because the ones at Dresden were completely destroyed during the war. Konrad is getting along rather fine, and when the weather will be more like spring—until now we only know coldness, snow, sleet and rain—I will take him out and get some pictures of him. You will wonder how grown-up he became to look the years since I took the pictures I sent you once. He naturally has to be taken especial care of since he won't get rid of the malicious morbific agents for some years. But it is now some easier to feed him well than formerly. Some time ago I managed to get a kid from a farmer; she has grown to a goat in the meantime and just a fortnight ago gave life to two lovely little kids. So we have the first milk of our own now from her since the times when our old she-goat died. We can manage to feed her since potatoes and beets are free and the prices not high, nevertheless. The milk is rather fat and good, so my mother plans to butter some. We have become rather like farmers, haven't we? Oh, the she-goat isn't the only animal on our farm, some chickens are also running about and killing the nerves of my mother for all their work of destruction done to the diligently planted beds of our small garden.

But we could bake a heavenly Easter-Cake and didn't need take pure love instead of eggs, milk and fats baked into it. Your fats are still a big help; since I only sometimes manage to get some from the farmers. Naturally the biggest help is your coffee and tea, because those items are nowhere to get.

About the CARE-package I get rather puzzled. There wasn't any sender written upon it; but I only could guess you as the person it came from. I wrote to the agency to inquire the name of the sender. Perhaps it was some friend of you being shown some of my letters.

Please, give my love to all who sometimes remember us; relatives, friends and pupils. God bless you all and keep you! Also He may give peace to the world and reveal himself to those who imagine a conception of the world without God. Please, pray for us.

Sincerely yours always

Christoph

P. S. I am glad you wrote the things you liked me to send. I often ponder what to select. Please, do tell me whether I should send you more of those flowers—I put one sample, in the last box—who are holders for little cards to be put on the dinner-table at a party to indicate where people are supposed to be seated. There are different holders and I would send them if they appeal to you.

When I took the last packages to the custom-office to be checked and sealed there before to be mailed I read a note informing people that now 25 cigarettes are allowed to be received in a parcel. Are cigarettes very expensive in the United States? If they aren't I would like to tell you that I in spite of not smoking sometimes met people who are better talked to when offered a good cigarette first. But it isn't really *necessary* to send them.

Christoph

This letter is interesting in that it describes how different life in Berlin is. The city is divided primarily by economics—West Berlin is expensive for East Germans; East Berlin is less so. Travel between the two is normal. In the rest of the country, it is difficult.

Christoph also tells of having placed an "illegitimate" baby whose mother died with a family only to have to remove her from that home after finding out it was unsuitable. Mary Fulbrook writes that 15.3% of births in 1947 were "illegitimate."

June 17, 1951

Dear Miss Grover

Dear Mrs. Blakely

About ten days ago two wonderful and precious boxes arrived safely. Many, many thanks for these riches which will enable us to go along for quite a while and make us fit for rendering the strenuous service laid upon us. Though the weather has changed completely driving around the country is still a great stress to the physical constitution, for instead of sleet and rain drenching the clothes to the skin now a very hot sun and a sultry air heavy with the dust of the roads tries to make the body outworn and tired; while until some weeks ago the water came from outward clothes now are wet the same caused by perspiration. I often am remembering the days of my studying-time when I could sit in more or less cool rooms and let the sun throw her beams upon those who had to walk outside. The trouble besides is that I am still a little weak from the long sickness, and the great energy I went around with last year is still awaited to return. But the doctor says I shouldn't get fidgety because everything needs its time and I should be thankful I had recovered that far already. I did make fine progress, and I am feeling it myself, since I am quite able again to concentrate my thoughts and even get to going on with my promotion-work. Also my professor of the New Testament was glad to see me and expressed even his hope to get a call for me to university. He and his colleague are now over 70 years of age and he would like to see the follower. And so he admonished me not to lessen in my private studying, though for the time being there seems no such opportunity and besides I couldn't let go back all the work upon the shoulders of my father.

Both of us are so thankful for your gifts, and as in winter the hot coffee and tea were a help never too highly appreciated we now changed mostly to iced tea and coffee and find everything delicious. Through your help we can render help to many people and I pray and hope there will be much blessings for life in this world and for the day of our Lord.

It is a law effective in creature that mind can operate more smoothly when the hindrance of the body is lessened, and though we hope for a time when there will be no need, suffering and tiredness, we are bound to the laws of nature as long as we walk in this body; even Paul had to complain when comparing the tasks he was called for with the possibilities given to him as a human being. I delayed thanking you for your gift because I waited dayly for letters of yours. We were very troubled about the long time you kept silent. The boxes were the first sign of you after nine weeks. Our prayers enclosed your healthiness constantly and now we are relieved, for the arrival of the boxes seems us to give evidence of you being well. How else could you have wrapped and mailed them? But also from Mrs. Blakely no letter arrived for a long time and I would be very happy to get a few lines telling me everybody is feeling fine.

We are always overwhelmed by your ever new readiness to help, for it gives evidence that it is really possible helping becoming a habit of life and character. It does good to know this, and it's the biggest help you are rendering to me to go on working regardless all the disappointments. Just in my work I so often experience enthusiastic promises and after a certain time I sorrily have to notice the emotions dying away. So some months ago a farmer promised me to take an illegitimate child into his family. It happened after a service and the farmer was so glad to do something in return to the love of God. And because I just had upon my mind the sorrows of a girl whose mother had died and who didn't even know the name of her father I thought happily to give her a family by this way. But last week we had to take her away from the farmer, it was like hell there.

But my life isn't made up just by a lace of dark pearls. So a newly wedded couple who had a seat of their car vacant invited me to take part in their honey-moon trip which was to be spent in Berlin. And since obligations were just running easy my father took them over and I went a few days to Berlin. The city is a place of unique historical importance and of a very strange situation. All four big powers have occupied the capital, each holding a forth part of it. But since the city can't be split up, there is to be found nothing like the zone-border which divides eastern and western Germany so making it nearly impossible to get to Germans living at the Rhine or in Bavaria. In Berlin it is different. You simply drive in a streetcar and mostly don't even notice in which sector of the capital you are going. Perhaps you might see at one station some Russian officers or some men of the police of the DDR and at the next station Americans may enter the train; then you naturally will have noticed that western part of the city is entered. People can go everywhere absolutely unhampered, some work in the sector belonging to the DDR and live in the American sector or the other way around. Naturally the going around is complicated by the existence of different currencies. While you might buy in this shop for money being valid in the DDR (Russian Zone) you perhaps might only cross the street or walk a few steps and enter the next shop where you are surprisedly asked for money valid in the western zones of Germany. Then you mostly have to leave the shop without having bought anything, because the exchange-rate of money is very high. So when boarding a streetcar in the American sector I have to pay the equivalent of the one-hour-pay of a skilled worker; and for the rent of one single room I would have to pay all my monthly salary. Those who work in the DDR sector therefore get some % of their DDR-money exchanged at the proportion of 1:1; but those who live in the DDR sector but work in the American one get only 20% American zone money. So everything is very complicated as you see, but very interesting as you might imagine. You simply go around and see the representatives of the four powers peacefully settled together, without any border between them. Airplanes from everywhere come to Berlin, so it is a big meeting-place of the whole world. Next month the evangelical churches of Germany will meet there, many hundred thousands of people from all zones of Germany will come together and discuss religious problems. They will demonstrate in whole Berlin that all are brothers and that no zone-border can divide them. I will go too. Special trains will be provided and the prices—else being rather high—are cut down effectively. Wouldn't you like to come over too? Many foreigners will visit us and it would be a wonderful occasion for us to meet. I do long so hard to see you once.

At Berlin I also saw an American picture with Ingrid Bergman starring (I guess you say so). It was the one according to Hemingway's *For Whom the Bell Tolls* and left a deep impression upon me. Did you ever see it? It is the story of an American fighting in Spain against Franco in the civil war among the guerrilla-troops blowing up bridges etc. Over the whole picture hovers the uncertain doom of the outcome. Without being told one is certain he must dye, and it is him for whom the bell will toll.

And so all his feelings and his love become him a depth which only is experienced before the dark door of the future. It is strange that most people have the moments living was worth for in the face of death. Things we else hurry by when entering our mind or being not noticed at all get a stress when this feeling accompanies them that there are the last moments of live, the last day, the last night. Or that it might be the last. I know it from the battlefields and the crouching in cellars when bombs were smashing the homes and the air was filled with howling and bursting and fire. How many people have found God in such situations. How many learnt to know the secret power of love under such circumstances. There it really happened that one died for his friend, not by chance, but willingly risking this sacrifice. There were mothers covering children with their bodies found dead while the child was living still, there were men running into flames and death to save the burning and buried ones, and there were soldiers crawling from shelter for a wounded enemy.

But also the small things seem to have become big. The features of a beloved one, some words spoken in the past, the dew on the leaves, the home and mother pictured in the mind.

All the trifle events, the way she put the cap upon my head when leaving for school, or the way she waved when the train left the station.

It is rather a truth that misery, need and danger of life opens the eyes for many things and render the experience of feelings which are in spite of their sadness that very deep and wonderful no one has drunken from the well of human life, who never was touched by them; while richness and safety so often makes level-thinking, taking everything for guaranteed and therefore killing the possibility of noticing the real value of things.

It is strange that there are always the little things winning importance and the big ones—so much chased after—being veiled with insignificance, when the shadows of death are thrown where one walks. One catches the breath and suddenly gets aware of so much sun in one's life, until then appreciated not sufficiently or even not at all.

The problem only consists whether this is a more real apprehension of reality and world or caused by a darkened intellect which in bright days—not over ruled by emotional coloring—shows that things are as we look at them in daylight and not when day and night are touching each other, in the last gleaming twilight. But I think we only can judge upon the grades of light when we know darkness; yea, we wouldn't even know there was light providing the existence of night.

Well, my thoughts were running along but I often wondered why pictures with so many thrilling happenings supposed to catch the breath of everybody didn't touch me at all, and there only need be two people speaking everyday things but being transparent to the inmost things of life and mankind, a soul being open to one's eyes, a scene showing real life, and my heart goes over to them. It touches when this sphere is entered and one feels what human beings are.

There are the small but deep things that matter. The things coming from heart; the look of the eyes which let notice the soul living behind.

So my stay in Berlin happened to become a big experience concerning many sides.

But there is still another joy to mention. I am uncle. Ehrenfried my brother the gardener is since a fortnight father of a nearly ten-pound girl. God may bless her and her parents.

There are many greetings to tell you. Ehrenfried and his wife, my brother John, Konrad who is feeling fine and whom I visited just yesterday, and many greetings from my parents!

Also the shop-keeper wants to give you her most thankful regards. So you have lots of friends in this country, and you just may come over and will be heartily welcomed by many remembering you often and thanking you for your kind deeds and big heart. The Lord may repay everything you did unto us!

Sincerely yours

Christoph

In this letter we find Konrad has an interesting backstory. His mother is Jewish, and a girl Christoph had fallen for (and who went on to marry a doctor when Christoph decided to change his major to theology) is Konrad's half-sister. For much of the Nazi regime, those who were in mixed marriages were ostracized and suffered economic consequences. Often they had to leave their homes and live away from "Aryans" in "Jew houses," as happened with diarist Victor Klemperer and his "Aryan" wife. In 1943 Nazis rounded up Jewish spouses of "Aryans" for deportation, but the protests that resulted caused the Nazi authorities to relent. The 2003 fact-based film Rosenstrasse *portrays that protest quite well. In the last few months of the war, efforts were made to deport Jews in mixed marriages to the east. Klemperer was ordered to report for deportation in Dresden on February 16, 1945. He was saved by the conflagration that engulfed Dresden on February 13-14. Ninety percent of Jews in mixed marriages survived the war. Christoph writes his family helped Jews fleeing concentration camps—he says no Jew evaded concentration camps or death without help. If true, his father had come a long way from his days as a "German Christian." Perhaps Christoph's father found himself in the same position as Harriet Beecher Stowe's character Tom Bird, the Ohio state senator who enthusiastically supports enforcing the Fugitive Slave Act in Ohio until he encounters a fugitive slave in his own home. And note Christoph (possibly pointedly) refers to Konrad's father as a Christian German.*

August 12, 1951

Dear Miss Grover,

I am rather put to shame by your always new readiness to help having been revealed again through those 3 wonderful and valuable boxes that arrived safely the day before yesterday. What money you must have spent upon them! They certainly mean a richness to you too, but what value they have to us! Sending them you make so many happy. And so I am not only put to shame by this very example of Christian living but I am getting also very envious of you being able to render so much happiness to so many people. You really are sustaining and supporting some families and mor more with those treasures which aren't only rare and scarce in my country but are things to most only known by their names since they can't be bought in any shop or at any place but get to us only by this way as greetings of help and love from a country being far away but often thought of. It is like a glimpse into a world of light and richness. Though we certainly don't forget over all this richness that your people have their troubles and needs too, the unpacking of your boxes always seems like a poor child casting a look into a bright palace and getting dazzled by this gleaming brightness. What words shall I find to thank you? I guess there are none.

But another big joy was prepared to me by your nice letter I got after a long time of waiting in vain. From my last letter you might guess how I am now relieved when having noticed that nothing was wrong with you but you only were delayed by work. That I can understand too well. Your letter I found when returning from the big conference of church-people at Berlin, but on account of me being away some time from my work a lot of it had been heaped-up in the meantime and having returned I had to take all the time catching up with everything. One can't get away from one's obligations for a while without having to do it afterwards. So I can't even say I put off answering you, but I never had the time just to start. This Sunday afternoon were the first hours I had off again and I ... slept through all of them. And now—being refreshed I will soon start to leave for an evening-service. Well, there are some minutes, nearly an hour, left and so I sat down to tell you about all this happiness you created among my people.

The acquaintance of mine asked me to tell you her thanks too; she wonders whether to go on sending angel-figurines or some different ones; so already the last time when sending the napkin-holders we put our means together to send you the whole set. Now she was able to get hold of another big angel. For Christmas she hopes to get a set of Mary and Joseph and the child in the manger. Things aren't to get very easily since a certain quantity is destined for export and the private shops get only merchandise when those requests have been met. Life isn't too easy for the shop-keepers too. But we are going on well in spite of some difficulties. It is better than years ago when we were lining up for a pound of potatoes; vegetables, potatoes and bread may be bought all quantity one likes.

Also Konrad is feeling fine, I will have opportunity to meet him in about a fortnight. We have invited him so he will be able to eat some fresh fruits of our garden. Trees are full of them this year, and this means a big help too; so we will have jam and not need to eat the dry bread; the price for the non-rationed sugar was lowered too; I told you it was 6 Marks, afterwards 4.50 and now 2.50 Marks; but it is still nearly five times as much as the one we get on the ration-tickets; nevertheless we buy a pound hence and then for it is needed.

Konrad will also be very happy about your presents. He always asks when I meet him whether there isn't anything for him I brought along; that much he is accustomed to get something when seeing me.

How I met him first? Well, perhaps I shouldn't talk about all those things connected with it, but I guess I might do it to you, so it will certainly make me a little sad and [meditative] again.

It was years ago, in the second term of me studying English. During one lesson I noticed a girl-student sitting beside me by chance and she rather "knocked me off my heels" (I guess I read this phrase somewhere); I couldn't concentrate myself upon the things the teacher was telling us but only was wondering how to get acquainted with her. So after the lesson I made up a story about me having missed some lessons and asked her to lend me her notes. Some months afterwards we were engaged. And she is the half-sister of Konrad. He was just born at that time; it was the worst time we had to live through after the war and—as you know—his father (the second husband of his mother) died of TB which claimed many casualties at that year on account of the bad nutrition. Konrad's mother is Jewish, so she and her daughter had a very bad life during the Nazi regime. [In Orthodox and Conservative Judaism, a child's Jewishness is determined by matrilineal descent.] My fiancée was a rather good-looking and intelligent girl and I wondered why she hadn't any boy-friend until one day when she told me that every time when meeting a boy she approached the topic of the Jews and then she always noticed the repulsion against this race; there is quite some of that feelings yet! And so she dropped everyone. When we were walking home from the lesson I told you about she also approached this problem and I—without knowing her to be Jewish—told her about the Jews, run away from concentration camp, we had hidden in our lodging from the police; and thus we became friends. Her mother remarried a Christian German lawyer right after the war; he had helped her during the years of persecution (no Jew, who wasn't helped, evaded concentration camp and death in Germany) and Konrad was born as their little baby, and all of us spared from our small rations to pull this weak child through the years of sickness and hunger. And you helped too! And I couldn't leave him when I was left. It was the time when I changed to Theology. Then an end was put to the happy times. Today she is married to a medical doctor. Well, I can't blame her; but it was my first and only love. Since that time I [have been] alone. Certainly there are a lot of girls here, too many even, on account of the losses of men at the battle-fields, but can I dare to take one along with me into an uncertain future. No, we have to keep watch alone. Well, that was that.

And now returning from the past to the present I can tell you a thing I am very happy about. There was a letter from Mrs. Blakely telling she is feeling better now. But nothing I did hear from Miss Jamison and Charlotte Dean. I wonder what happened to them.

About the tea it is difficult to say, because the Tetley-tea isn't always the same quality. Some packages certainly seem to be better than the TenderLeaf, but others aren't. But *good* all this tea is, and you may send the kind you get easiest.

The little daughter of my brother Ehrenfried isn't only a joy to him but also to my father, the grand-dad; I enclose some snaps I took about 4 or 5 weeks ago. I hope you like them.

Many greetings from all of us and to all of you!

Sincerely yours,

Christoph

Christoph opens up about the emotional toll being a minister is taking on him. He is happy to have Ms. Glover as a sounding board, since he does not feel he can display any sense of weakness to those he's serving and certainly doesn't feel he can open up to his father. He's still a bit hurt by his failed romance, but he notes that people fall in love, and love doesn't last when real life intrudes. Christoph is twenty-six.

Looking at the romance from Renate's point of view, it's easy for me to understand why she broke it off. Christoph's decision to major in theology pointed to a future for her of being a pastor's wife in a country that discouraged religion. Further, she, like almost everyone in East Germany, was living in poverty. Who could blame her for choosing a doctor? Finally, Renate's mother was Jewish. How appealing could she find being the wife of a pastor in a religion that did not accept her mother and the daughter-in-law of a pastor who had been part of a movement that had sought to exclude non-Aryans from its membership?

Christoph also notes his older brother is having trouble with authorities over his farm's production.

November 18, 1951

Dear Miss Grover,

Work was very hard for me during the last days and weeks so it would happen that I postponed answering your letter until now. It's the time of the year when most of the burials are, and since they are on different places with drives of one hour between them a day easily is filled by them. At those times I really have to concentrate myself upon my conscience, so preaching doesn't start to be a mere matter of routine; the danger is near to all professions where the heart is easily taken out from them and all is going smoothly like a well-oiled machine, but it nowhere seems me to be that dangerous as in the professions, both of us are in. I feel it the same in the religious lessons I render. The individuals with their different questions and problems, their hearts and souls, which are little worlds of their own, encountering new difficulties. I since long had passed by, are forgotten easily and I often find myself overcoming everything with an experience of mine, trying to solve a matter without going through all the problems now, living through them again, as a real leader should be. And so it is not difficult to forget that a deceased one isn't—for those I am adrdessing—just a member among many as it might be to me, but only in a moment when I have lost the right to preach anyway. So it needs so much praying and trying to feel the hearts of the others I am outworn each night and often notice in the morning I even had forgotten to switch off my lamp by which I had tried to read a little yet. But I am thankful, I can be outworn yet. It is the sign to me I am not drowned in routine yet.

Please, don't be angry with me about me talking about all this, but I have to write or tell things sometimes I couldn't confide to anybody hereabout. For you might understand talking that way if it became public would undermine the trust of the people upon their parson. Most are plain thinkers and can imagine everything only to be this way or that other; a living between, and a constant striving for the realization of the aim they won't grab. And others will think: If he already concedes that much, it will be a lot worse with him. So I haven't anyone to talk things over the way I would like; because I also don't dare to do it with my father because he really seems to stand on a higher level than I am, and for him things really seem to be natural I always have to fight for. So I am glad I know somebody I can talk to straight and upright, as I only can do in my prayers besides. The friendship of you and Mrs. Blakely means a lot to me, not easily to be explained by words; it seems me to be more an undercurrent of feeling which combines us. Wasn't it strange that you already felt the things I talked about in my last letter, though I anxiously tried never to let drop a hint. It's still a sore which breaks up in hours of loneliness, which means nearly all the nights sitting in my room and listening to the wind singing in the chimney. I certainly am not more in love with Renate (that's the name of the girl), but it's the remembrance of a time I still believed in marriage and happiness, that makes me feeling bad. I was that sure all was going right on living at two, so in this disappointment all my faith was spoilt. And this mistrust was reassured nearly each time I meet couples married by me some time ago. When they announced their wedding, their features were radiating with happiness and all their hearts lay open within the gleaming of their eyes, but when I meet them later on, perhaps at occasion of the birth of their first child, I notice this light having gone away from their eyes. They try to act happily (what should I think otherwise!) but if I can get them to confide to me they concede they are staying together only for the reason of the child or for the opinion of people because he is in a profession whose personal life is looked upon; others even talk about divorce, and all this after some months! I wouldn't be that shocked if I hadn't seen them before really happy, because they really *were* it, they didn't only *believe*, they were in love, no, they really were. So there seems to be something which can't be handled by human beings themselves; it's given and taken away; I don't know by whom. Certainly it can't be done by God; for he wouldn't create such hells I have looked in. There are other powers ruling in the world besides. People seem to be not able to do anything to it. Couples may stay together for the sake of duty or friendship, but the real love is beyond their disposal. And since most people are without any sense of duty, you might imagine what chaos will be created after the love, the only link between them, is broken. So I stay alone in spite of the long hours of loneliness. I heat my stove and I prepare most of my meals. I am happy I am near home, so my mother cleans the room and washes my things. There are girls falling in love with me and I come to the same feeling sometimes too, but I pass it unnoticed. What should people say, if the minister's marriage wasn't an example and how could I go on preaching to others upon a dark background? I sometimes wonder why you stayed alone.

Now there are a lot of questions to be answered, you put in your letter.

It wasn't easy for my brother Ehrenfried this year either. You know they started this little vegetable farm without any implements and animals and you know there is a certain amount of foodstuffs to be delivered. The quantity is settled according to the bigness of the farm. For each acre there has to be delivered a certain amount of different vegetables, potatoes, but also eggs and milk. So he is supposed to deliver milk, but has only calves. And there are different troubles of similar kind. Since the administration-officers didn't take any notice of the special situation of a newly built-up vegetable farm, my father went with Ehrenfried to Berlin to our government. Now we have great hopes conditions for him will improve. It often happened already that people were successful by this way. His little daughter's name is "Dorothe"; she is growing steadily.

All the others feel quite well. My brother John had another fit this summer but it is over now and he enjoys his work again. Father and Mother are feeling fine, too, and I myself am over that flu since over half a year, I think. I hope I will pull through this winter without such mischief. I will try to take more care of myself.

Very troubled I am already by the approaching Christmas time with its scores of celebrations. Celebrations in each village; and each small group of a village has a special one besides. And I don't want to talk the same everywhere. Please, do pray for me, I will have all the spiritual help needed for this task.

I hope you will have got my Christmas-package in the meantime. I managed to get two different sets of Mary and Joseph. The one is from me and the other one from my father. But the wishes from the members of my family accompany them. May the Lord give you a blessed Christmas and let you feel the importance of this day not only for the whole world but also for each single person, and may he guide you safely through the New Year and render you the needed strength for your work at the new place you are at. And may he—I also dare to pray for this—keep our friendship going on.

Since your letter I didn't send any package from my shop-acquaintance any more. I even was a little glad you suggested to drop it for a while, because I was a little disappointed by her. You see it was a long time since your last parcel arrived and she always were a little doubtful. I told her about all the difficulties you had in sending them and I assured her, if they wouldn't arrive I would pay her loss.

I never had any doubt in you remembering us, and I was that happy I was right. The one package with the cross-mark already has arrived. So I put her to rather a shame. I don't know whether you feel what big help it is to me to have evidence that there *are* people in this world who are quite otherwise than those who go around without any love and faith. Evidence, that there aren't only words of preaching to be listened to, but facts to verify them.

I hope you got the napkin-rings in the meantime. If you didn't, please do write me, because then I have to inquire after it because I sent it by registered mail.

Well, now I have to close, since it is already very, very late. The morning will arrive soon and it is the time you will start to sleep, because the sun left you 6 hours later than me. Well, it will be soon with me again, but I have to try to catch some sleep of a few hours before, otherwise I won't be able to have my lessons and addresses. I don't hope you will notice from this letter that it was written after midnight.

Sincerely yours,

Christoph

P.S. Some days ago I wrote to Mrs. Blakely. But there was no letter from the niece of your friend.

Christoph notes a package was delayed by a strike in Hamburg, which was in West Germany.

November 19, 1951

Dear Miss Grover!

Today the second package arrived. It was delayed by the strike of the German dock-workers at Hamburg. Many, many thanks for this precious gift and all your troubles you had by buying, packing and mailing. I know it needs quite some time and energy. Tomorrow my brother has to go to Leipzig. I will ask him to take some things to Konrad, Yours truly, Chris.

Christoph has good news! He has been accepted into a doctoral program at the University of Leipzig. He almost turned it down because he was needed so desperately in his old position, but his father pointed out teaching those who would become ministers is also important work. The professor who supported Christoph's entry into the doctorate program is almost certainly Emil Fuchs. Christoph also finds a new roommate and sees A Streetcar Named Desire.

February 2, 1952

Dear Miss Grover,

I received both of your parcels last week and your very dear letter already quite some time ago and though I welcomed and appreciated it with real joy I couldn't give way to my longing to answer it at once. It just arrived at the New Year's Day, and since this New Year's Day meant for me a stepping-stone onward and a change into quite a different world of obligations from that I lived in during the last 18 months, I was that happy your greetings and good wishes and all those words of friendship were accompanying me into this year which—as it turned out—put me into more difficult work than before. Some days before Christmas I got a letter by which I got a call as Doktoraspirant for Professorship of the New Testament at the University of Leipzig, Theological Seminary. I guess I already told you formerly that my professor wanted to get such call for me. I never believed it would happen since mostly older ones get this chance. Therefore I had myself settled down in the meantime and by the rush of all my work had nearly dropped the idea to go on with any scientifical work, and forgot all about becoming a teacher at the Theological Seminary. Now it has happened I was quite surprised and at first didn't feel like going. How could I leave my other job I loved that dear and all the people in whose sorrows and joys I had taken such partnership. On account of the shortage of ministers I told myself I had to stay at a parsonage. And all this just at Christmas time. All my fears of having to overwork myself and be troubled by the many addresses I had to deliver were in vain and everything was filled with more grace by the Lord than last year. So you may understand that I could have made up my mind not so easily. But talking with my father changed my mind. I was made understood the even more urgent necessity to educate those who want to become ministers. And so I left. I didn't leave absolutely, though. Sometimes I go to Dresden in order to preach and meet the congregation, but there is not left too much time for that. Being a teacher you know that the start needs the most preparation work, while later on the constant repetition of at least part of the matter might ease the strain of work a little. I have to catch up with so many things at once that I hardly know where to start. Within the next years I have to finish my promotion and my habilitation work. I don't know whether I will have the strength to live up to the expectations set upon me or to give up. But I know yours and Mrs. Blakely's help will have a part in the eventual success.

It happened just on one of the first nights I stayed at

Leipzig that I found somebody which will be now of company to me in the single room being the new home of mine. I returned from Seminary very late and had to pass some very lonely and dark streets when I noticed the pitiful crying of a small kitten. At first I thought she was sitting in front of a door which lead to her "basket" and so I made up my mind to go there and ring the bell. But when I approached, I saw neither house nor human beings but only an about one-month-old kitten shaken by the snowy winter storm. She was that meagre and weak that there was no doubt left about her being put out by some people who didn't want to keep all the children of their cat. One says the little ones born at winter-time don't look nice and have no sleep pelt. But in the cry for help there certainly is no difference and this I must have felt. And so I put her under my overcoat and took her home where I had trouble to find something she would eat. Well, there certainly isn't anything in such incident like that. We pass and have to pass so many sufferings that it seems childish to take regard of output animals or even to find them important enough to write about them. But there seemed to me sufficient symbolical meaning in it to take it more important than usual. There is such a lot of "freezing" and crying for help in this world, there are so many "lost" ones, but there is never lost the chance to get found; never there is reason for complete despair as long as one may suffer, and for all troubles may be someone coming around the corner to heal them; perhaps a human being, perhaps God. So the little cat—I called her "Foundi"—is a good example to those many who also here in Leipzig knock at my door to talk about their troubles, troubles of quite different kind than the little animal had which only needed some warmth and food. It's that simple to do something in this concern, but it is that very difficult to spend warmth and food for the human souls.

Now "Foundi" seemed to have noticed that I was writing

about her, because she stopped sleeping and jumped upon my desk to look at the strange lines of ink pouring from the stick in my hand.

There was another event when I had the opportunity to see the American picture, *A Streetcar Named Desire*. It's a movie which doesn't lift one's spirit but I left it rather smashed and felt depressed the whole day afterwards. It is a problem which might be discussed whether art has the obligation only to give high aims and happy ends. I think I learnt a lot of this picture which was of quite opposite kind, nevertheless. A girl having given herself away to men all her life wants to have a new start and find a fulfilling marriage. But it is too late now, and so her desire was crushed. Nobody wanted her for this concern any more on account of her past. It's strange that in the hearts of those people who have fallen deepest, the cry for going home is growing more and more urgent. I often had to notice it, and I think Jesus must have been touched by it when he tolerated his feet to be kissed by the woman at Luke 2:38.

But among the books I was reading during the last time there was an American novel which was of the uplifting kind. *A Bell for Adano* by John Hersey. If you don't know it yet, you certainly have to read it.

The weather is not that cold any more, so the moving and staying here isn't too much hampered by cold degrees. But while giving up that room in Dresden I must have put a letter I got from your friend into some book of mine and can't find it now when wanting to answer it. I am very troubled for this reason. Please, would you give her my greetings and excuse the delay, and please do write me her address next time.

For Ehrenfried there is no definite answer from Berlin yet. Until then the most urgent things are suspended. Some officials are working very slow. But we hope for the best.

Yesterday I visited Konrad. He feels very well and wants

to give you his greetings.

All my family is thanking you for everything.

Most sincerely,

Christoph

There are new restrictions on what can be sent into East Germany, so Christoph asks Ms. Grover to send packages to four addresses.

February 5, 1952

Dear Miss Grover:

Since I just got acquainted with the new restrictions of import I will write another letter beside the one I send by normal way today, I will give this one for air-mail, wondering whether it will go quicker;

Now the restrictions are the following ones:

Each person may receive:

2 pounds of ground coffee (or instant coffee which is weighted the same as ground coffee, so this is of profit!)

or 3 pounds of green coffee

and 2 pounds of cocoa

and 125 grams of tea—for *each* month.

The other items are not restricted (only those things which aren't to be bought here!). So I would like to ask you the favour to split the two parcels into four, if that isn't too inconvenient to you, and to send them to different names.

The one with the cross (now being two ones) to:

1. Pfarrer Ehrenfried Haufe

   Dresden A29, Meissner-Land-Str. 30, etc.

2. Johannes Haufe

Dresden A29, etc.

and the other one to:

3. Mrs. Edith Haufe

Dresden A29, etc.

4. Christoph Haufe

Leipzig W31 Pistorisstr. 43b

Germany, Russian Zone, D. D. R.

To Leipzig you may send more tea, because there doesn't seem to be this restriction.

I know it will mean more trouble to you, but you see, though I don't agree with this legislative direction I have to obey them. Please, do understand me, and don't let those directions be a reason for disturbing our connection.

Sincerely yours,

Christoph

Christoph, and probably everyone else in East Germany, is confused about the new restrictions.

March 6, 1952

Dear Miss Grover:

When being home for the birthday of my brother Ehrenfried your wonderful parcels arrived completely safe and were delivered to me. So the restrictions announced by the custom-office some time ago—those I wrote you about in my last letter—aren't valid yet. I was very happy about that as you may imagine. Some people say those restrictions won't ever be in validity, they only were issued so Americans would get informed not to send that much of those things that aren't available here—on account of propaganda. You see, we are far better off now than years ago; there is sufficient bread, potatoes, vegetables, fruits and even sugar; and so when people get told they should be content about this big progress, they hint at those items they ain't able to buy but which might be bought in other countries.

But I don't think it's only a fake. They may as well just wait that long with those restrictions until all parcels, sent off before any announcement could have got to the sender, could be handed out. The reason for the restrictions might be the black-market they want to stop. But naturally I don't know exactly. Anyway, your parcels are split up in that many parts and help that many people, so there was a little worrying about those restrictions. Each person may receive 2 pounds of coffee and 125 grams of tea each month. So I asked you to send the parcels to the name of my father (Ehrenfried), mother (Edith), and me.

Sincerely yours,

Thankfully

Christoph

Christoph is settling in and seems to be enjoying his new position—even fashioning what must be a very rough draft into a book for a "famous scientist." The book is Emil Fuchs' Marxismus und Christentum *(Marxism and Christianity).*

May 3, 1952

Dear Miss Grover,

Having been home over the Easter-Holidays I received three parcels of yours arriving safely just at this festival. There was a big Hallo from all of us when the precious and most welcome things were spread upon the table. But there was even more joy waiting for me. Having returned to Leipzig I was surprised by two other parcels some days later. Many, many thanks for all the trouble you had by shopping, packing and mailing. I know it takes a lot of time. And you do all the things by yourself, while I am helped by my mother, so the work is split. Now the custom office man told me that the restrictions I wrote you about were lifted again. So the other reason I told you about was the true one. Nevertheless I would like to ask you to write different names of my family on the addresses. But you may now put in any coffee or tea you like. Perhaps you don't have to pack that many parcels, and you could send the things you intend for Konrad and me to Leipzig, and the items you intend for my parents and brothers to my father; and if you want to send some more to the shopkeeper, you may take the name of my brother or mother. But there is no need of doing it, and may do it as formerly.

Konrad will enter school this year (summer-time), so he soon will be able to write you a note, though it will be in German. But soon I will start to teach him English. As happy and thankful we were to receive your parcels, as worried I am because no note of yours arrived, least a letter. But I hope it already will be on its way.

My work means now a hard task to me. Though I have only to render just two lectures for beginners, it needs quite some preparation-hours; I have to read a lot, and so I often sit until late in libraries or in my nice room. I also can go on with my promotion work, so I don't have the time to work for it directly; but since I have to study that many books which touch those problems too, the heap of notes is growing higher and higher. Besides I have to correct and overwork a manuscript of my professor intended to be published. Though he is a famous scientist he is a poor writer nevertheless. He starts sentences without finishing them; he jumps from one idea to another without giving an understandable connection; so I have to polish up the style, and that's a difficult task, more difficult than writing a new book, because I have to ponder over each sentence. But now I have to close, because I want to air-mail this letter. I will write another soon,

Sincerely yours thankfully,

Christoph

Here we learn more about what is expected of Christoph with regards to the book he's reworking.

A quick search of the internet reveals Christoph and Emil Fuchs became close, and Christoph participated in other papers by and about Fuchs.

We also learn the import restrictions never went into effect, and Christoph believes they were probably a response to American propaganda.

Christoph's brother was successful in his efforts not to have to provide milk and meat from his farm, which had only calves. Sadly, Christoph's kitten died.

21st of June, 1952

My dear friends,

dear Mrs. Blakely and dear Miss Grover,

It was a wonderful day when about three weeks ago the letter of Miss Grover was received by me. I at once started to answer it but got along only a few lines until the bell was ringing and my professor showed up with a heap of new pages of his manuscript to be revised by me. And there was no proceeding of the letter-writing since he didn't stay only a few minutes but started a discussion about the problems moving his mind and—as I am certain of—also his heart, so I would be able to mold them into a form not being opposite or inadequate to his intentions. I already wrote about this manuscript of a professor who is rather well known on account of his scientific research.

In our country each book which shall be published has to get a license before. If it is rejected then it cannot be printed. Quite some manuscripts are declined on account of the paper-shortage or for other reasons. But he managed to get a license and went happily to the publishing-house. Some days later he asked me to accompany him to the publisher and there I was told that the book couldn't be printed before the style and ductus was rearranged. Though a rather clever man, he is accustomed to write as he speaks and so it happens that there are sentences started but not concluded or in a wrong way continued. But this was not the greatest difficulty to be overcome. Since he knows the suppositions for the understanding of his written thoughts he sometimes jumps from this thought to another without giving a connection being necessary for the understanding of any reader. Now he was neither in the mood nor did he have the free time to retouch the whole manuscript and so I was asked to do it. There was no other way for me than to say yes, though I sometimes felt sorry afterwards. It was a more difficult work than to write something of my own. Because first I had to feel myself into his way of thinking, then I had to change for hundreds of pages all the sentences in a way, which still was somewhat like his writing and not my own, and at last I had to compose quite a number of sections to connect his paragraphs. But in spite of all it was quite an interesting work to do, though I didn't [conform] with some ideas of him. The subject was the social gospel of the prophecy and the New Testament, and those compared with social ideas of non-Christian socialists. Most of all the stress was laid upon the fact, that Christianity is not—as often reproached by others but also believed by many Christians themselves—a congregation of just religious people whose faith has no connection with the way the world is going on, but is supposed to be the salt of the world and try to realize the commandments of the Lord in this world to make it a better one, a world of justice and without hunger and need. "Repentence" and "Justification" are no means to rely upon without the constant struggle to live up to the way Jesus showed us. Only on account of the fact that the churches of the 19th century and before and afterwards neglected and forgot their task to have a part in shaping the world according to the will of the Lord and not according to the will of rulers and monarchs the non-Christian social movement was possible. And now there is shown and proved in this book, that—because this movement overtook only part of the real Christian gospel—the overtaken part is suffering under the lack of the other one and many mistakes originate in it. If people want to build a real social world they should find back to the gospel of love and to the faith unto the heavenly Lord, not to trust in arms and air-planes but in the all-pervading spirit which knows no hate but conquers the world with love. He is a Quaker and Pacifist and so he doesn't believe in revolution and war. There wouldn't be no need of both if there would be more of real Christian spirit in this world; and they only were possible and even necessary on account of the lack of it.

There is in this book a lot of material from the Old and New Testament and from German history.

Yesterday I finished the last page and so I may start now to think about other obligations which had had to be put off until now; not only on account of the time for this work but mostly because of the constant meditating of its problems much else didn't enter my head. So it comes that I wrote such a lot about it in my letter now.

The other task of mine was finished last month with the end of the spring-term and starts in September with the fall-term again. In the meantime, I am allowed to take 4 weeks vacation while during the other weeks I have to work upon my promotion and to prepare myself for the lectures of the next term. I don't know yet how many "classes" I will have, it depends upon the number of new-enlisted students. At university we anyway don't have "classes" like at school; but there are different lectures and "classes" announced and the students attend them they have selected by their free will. They only have to prove at the final examination that they visited a certain number of them. In the lectures we only have to deliver addresses while in the "classes" we have discussions with the students about various problems. Since I am not a professor yet—and it will take quite a long way to this title—I only have to introduce the new students into some principal problems of studying and Theology, especially of the New Testament; there is the teaching of the Greek language and the reading of the Greek New Testament. While reading we talk about some problems we encounter but naturally not with the profundity with which it will be done later in the classes for older students. Nevertheless there is quite some preparation-work to be done, because I don't want to teach only that I just have read but want to feel myself at home within the problems talked about. There will be no more than 10 hours each week I have to teach.

Now this certainly doesn't leave sufficient time to do a regular service at my churches near Dresden. But when I visit my parents over weekends—it doesn't happen each week but at least once each month—I go to a church to deliver a sermon. I like to do it in Leipzig too, because I don't want to forget all about the practical service. This isn't too strenuous. The tasks which took up most of my time formerly were not the Sunday-services but the many ceremonies in church and on the cemetery, and besides the many private problems I was asked to counsel. There were broken marriages, unhappy orphans, etc., I remember I told you about the child I tried to find parents for. All those sorrows were a now not experienced stress not only upon my time but far more upon my mind. I couldn't feel much happy any more with the knowledge of such a lot of suffering.

My father is still doing the superintendent's work, he never seems to get older and weaker; it is as if some spirit of him puts a lot of energy into his body. He always asks me to translate your letters, and when I did read to him about those 4 years of our connection he cried: And what sacrifices you have brought in this time for our sake!

In my last air-mailed letter to Miss Grover I told already about the safe arrival of your boxes. I hope this letter came to you in the same quick time as the other one. We can now write any air-mailed letters. Your letters take different time to reach me. From one to four weeks [in my] experience. The last one was seven days on its way over the ocean. I wished I could travel that fast and come on a visit during my vacation.

The restrictions for import are now lifted and cancelled again before they were valid at all. I already had a hunch about it when I wrote you; and a reason for it were the different ways those restrictions were proclaimed in different cities. Well, I am and we all are glad about it, because your boxes are more than something to relish but means of efficient help for our work.

My brother Ehrenfried still has to work more than all of us in order to bring his garden to a standard being able to support his family and to meet the claims put upon it by the agricultural department. He gets up at 4 each morning. But his drive to Berlin was a successful one. I wrote you that he was supposed to deliver milk without having cows, because the quality and quantity of food a farm or garden has to deliver is not measured according to their animals, etc., but to the [artificial] measure. Since Ehrenfried started with two calves he couldn't meet the claims for milk and meat. But his visit at the agricultural department in Berlin was of success, so he now can go on without the troubling sorrows. He built a small green-house this winter and so had quite some success in growing early vegetables. But all his tomatoes and vegetables he had planted outdoors were frozen by the wave of coldness which passed Germany a month ago. Also apples, plums and cherries are frozen.

My kitten "Foundi" died some weeks ago, she couldn't eat anything any more and so she grew weaker and weaker. My hostess who first had told me "I hate cats!" so that I had been very much afraid to take Foundi home with me in the meanwhile grew that fond of this little darling that she cried a whole day over her death. She—being a lonely woman of 52 years—had got that accustomed to the play and running about of Foundi that she only was consoled a little bit when I got a new kitten for her. For me it was not without interest to see, how also in these people who hate animals a deep love towards this creature can be awakened (after 52 years!). It reminds a little of the problem I talked over when mentioning the picture *A Streetcar Named Desire*. I couldn't imagine anyone being mere feeling and soft with an animal than her.

With very great sorrows I follow the political developments which make the rent going through Germany deeper and deeper. It is now very difficult to obtain a passport to cross it, since western Germany joined the Europe-Union.

Very troubled I also was about the storms and big waters in America I heard about. I saw some pictures of whole towns being covered by water; and I hope and pray there will be no harm to human beings and that you may not be touched by this calamity.

I am very troubled too by the long time which elapsed since the last letter of Mrs. Blakely. May the Lord give you healthiness and strength!

Sincerely yours

Christoph

Christoph is taking advantage of his summer break. This is possibly the first leisure time he has experienced.

POST CARD

July 8, 1952

Dear Miss Grover:

Many greetings from the "Baltic Sea," or as we Germans call it, the "Ostsee." I spent here a fortnight vacation at the fisherman's village, "Heringsdorf," which is situated near the Polish border on the Island, "Usedom." The weather is wonderful, and I am already dark-brown. The days are spent with swimming, sailing, promenading, and listening to concerts. It's a wonderful sight to walk along the seaside when the sun is arising and the waves are splashing against the land. It does me a lot of good, not to sit before books in a room, but be always in the open air. I am only sorry the days pass that quickly.

Sincerely yours,

Chris

Christoph expresses concern that theological seminaries may be abolished in state universities. He also notes that there are no restrictions on what can be sent to East Germany—for the time being.

August 23, 1952

Dear Mrs. Blakely,

Dear Miss Grover,

This morning I received a letter from you, dear Mrs. Blakely, and was very happy and thankful to hear about you getting better and being able to go to church again. From my many visits I know the feelings of those people who wanted that very hard to attend a service surrounded by all the crowd of the same faith praying and singing with them, but had to stay home and wait for me paying them a visit. I was deeply relieved to hear from you again, because on account of the long time since your last letter I was rather troubled. There is really a reason for thankfulness. Always when ministering a service and we come to the silent prayer I remember both of you and enclose you in all my own sorrows and pains I render to the Lord. And I certainly believe in the acceptance of prayers, though they naturally are no guarantee for the fulfillment of our own wishes; but if there would be no possibility to get God to change his will or to take care of something he had let be to itself, why then should Jesus have told us to pray? So I may see in your improvement a further assurance of my faith and my hope to see you face by face some day.

Sometimes I feel sorry to have no parsonage. So you see, I am like you partly sorry and partly glad. All those things which have worn me out formerly and I was a little glad of to have got rid of, I now start to miss. It certainly is good to study and teach Theology and go to preach at some churches about each fortnight, but it was different to be in the middle of the trouble of down-hearted or despairing people having dropped their every-days mask and looking with eyes through which the cry for help is clearly visible. It certainly is terrible to be often there with no possibility of material help and so to be reminded each minute of the day of this or that misery covered by a wall of this or that house I pass, I felt in a strange way happy about, all those hearts which were opened to me. And it never happened—as you refer to—that people did think: "He says it, because he is a minister and gets paid for," since everybody knows, a young minister here gets such low salary that he hasn't even the money to pay for a pair of shoes or trousers. In spite of having studied being poorer than many of them is really a reason that often takes down the dividing wall which often was between minister and people. Perhaps I will return to ministry sooner than I imagine, because I read, that the Theological seminaries of the State-Universities probably will be abolished.

I am now passing the second part of my vacations at home. The first fortnight of my four weeks I spent at the Baltic Sea. We had the most wonderful weather of this year—it never was that good again—and some students I was with watched anxiously over me not to touch a book once, because this was the wish of my professor. So I had just to lie in the sun or swim as far as the waves would allow us to go with our plain human strength. Sometimes we went sailing and during the evening-hours we listened to the murmur of the sea and when cutting the meditating silence talked about various problems, or we went to the concert or some other performance. This rest certainly did me a lot of good and I returned dark like an Indian. I took up my work and went on rather smoothly; on Sundays I preached at Leipzig enabling some ministers to take a furlough too. Also the university owns a big church where the professors of the Theological Seminary are preaching; I had to do it once too, though I like more to go to plainer people in order not to lose connection with them and get accustomed too much to a way of speaking only understandable to studied ones. And the services at university church mostly attended by students or academically educated people, so a sermon doesn't much differ from a lecture I have to deliver anyway during the week. But I will be obliged to preach there later, regularly, now it is only a representation for professors.

At the 19th of August I started off to Dresden to be there with my family at my father's birthday and stay there until the one of my mother at September 1st. The first celebration has passed by. My father has been a minister now for 30 years at the same church and so there were a lot of congratulations. You see, he often baptizes now the children of those he has baptized too; and between these dates there is a whole life during which they more or less were in connection with their minister. After the rush of congratulators, we were together at the afternoon-coffee, my parents, brothers and some visitors and we had wonderful hours. Most of all the little grandchild was a joy to all being able to walk now from person to person and ride on the arms of granddaddy through the garden. My parents now start very often to talk about the time when I was that young and tried to speak the first word and always tried to touch everything. Also the material servings were delicious. Skillfully my mother had baked a chocolate pie, and naturally we had real coffee, all things we wouldn't have without your help. So we talked a lot about you staying in such true friendship to us during these years. Your last 4 parcels had safely arrived some time ago. And we don't know how to thank you. There are no restrictions at all any more at the custom-office.

During the days staying at home I try to help my father; so I deliver some speeches about scientific topics and theological problems which some people had asked my father to get explained. But I also visit the other villages near Dresden. So the days pass very quickly. On the second of September I return to Leipzig. I have there some snaps from the seaside which I will send you.

During these two months I sent you 3 packages. The contents of two were table-card holders; to be placed upon your table before the seat where you want a certain person to be seated. In the socle (pedestal) upon which the insects and flowers are standing there is a little slit to put in a card with a name. The contents of the third to Mrs. Blakely are napkin-holders to keep the napkins of the persons you invited for dinner within the ring upon which the little figurines are placed. Before I had sent you some vases to put in flowers. I hope everything arrives safely, the glass of the vases and the little flowers with their frail and fragile stems (stalks). My mother and I tried to pack them the best way we could, because we thought they would look nice upon your table and be a joy to all visitors of yours.

With the best wishes and greetings from my parents, brothers, Konrad and me,

Sincerely yours,

Christoph

August 26, 1952

Dear Miss Grover:

Last Saturday I mailed letters to you and Mrs. Blakely and today two parcels arrived here at Dresden. So I want to inform you at once this time about their safe arrival. Many thanks for all those riches which came to us over the ocean again. Mother's birthday will be a good celebration of our family on account of them. I shall give you the thanks of everyone, also of the shop keeper.

I enclose a copy of the letter of the day before yesterday and will air-mail it. I am curious how much the difference of the arrivals of those letters will be.

From the stamp upon the parcels, I noticed that you are at Hendersonville again and I wish you some more nice days of relaxation. Please, do give my greetings to all of your family.

Sincerely yours,

Christoph

Christoph reveals that he disagrees with those who are so strictly religious that they oppose watching films and enjoying newer music. And this is before rock and roll! Evidently American plays are prohibited in East Germany, but British plays are not.

I think the most interesting thing Christoph discusses in this letter is having seen the 1949 film Lost Boundaries. *I had never heard of the film, so I watched it.*

It's based on a true story of a light-skinned Black couple. Both have degrees, and the husband is a doctor. In 1922 they move north and are advised by Black friends, the wife's light-skinned Black father, and a white doctor who has witnessed the husband's work to pass as white. They settle in a New England community (the real events took place in Keene, New Hampshire) and are well respected. In 1942, when the doctor applies for a post with the U.S. Navy, their charade came to an end. The Navy discovers from the college the doctor attended and a fraternity he was a member of that he has "Black blood in his veins," and he's pronounced medically unfit for duty. The film, which I'm not sure could even be called a B movie, makes some good points about how race can change people's perception of people they've known for decades. The film ends with a church service in which the family appears to be embraced by the community. The Keene, New Hampshire Sentinel, in an article commemorating the seventieth anniversary of the film's release, reports the doctor continued work-

ing until 1953, when he lost the job. He continued seeing patients in his home until the mid-1960s, when the family moved to Hawaii.

The film was controversial when it was released mainly because Mel Ferer and Beatrice Pearson, the actors portraying the couple, were white. The story behind the film was written in 1947 by W. L. White, who wrote They Were Expendable *in 1942 and condensed and printed in the notoriously conservative Reader's Digest in 1947, which may have helped it avoid scrutiny by the House Un-American Activities Committee.*

It would be very interesting to know what reaction Christoph's benefactors, who lived in the South, thought about Christoph's view of the film, which was banned in Atlanta and Memphis.

September 13, 1952

My dear friends,

Having returned to Leipzig I found the long letter of Miss Grover and two parcel-cards on my table waiting for me. I at once went to the custom-house to fetch them and was envied by a lot of people when carrying home the English-printed parcels. Everything arrived in good condition; never there was any glass-container of Borden's coffee broken; so I had thought I could send you the lovely vases of glass. I am sorry they were smashed to pieces.

Again there is a lot of reason to write you and thank you at once. Riches were really showering over me like in the fairy-tale telling of the little poor girl into whose lap golden stars had dropped from heaven. Though the standard of living has arisen since the former years when people had been eating potato peels, the coffee, tea and cocoa you sent are as precious rarities as ever before; they aren't available yet. I told you already about the birthday celebration of my father. If we compare it with one of 4 or 5 years ago we really have to be content. There was meat on the table and cake in the afternoon. But the coffee and the chocolate-pie (tart) and the tea at night we have been able to serve only for your kindness. At the birthday of my mother, September first, the plums just started to turn into the blue tint, so my mother baked one of her famous plum-cakes we know that well from the first remembered moments of our childhood. It was wonderful weather, so all of us were seated around a table put under a big lime-tree of our garden. We all felt happy and only were sorry that you could not be among the visitors. You could have got acquainted with all the members of my family at once, because everybody was assembled under the lime-tree talking about the experienced joys and sorrows. I had to translate the letter of Mrs. Blakely which had arrived a few days ago, and all were glad and thankful that it contained good news. How otherwise could we have contently and happily drunk the coffee whose donator we might imagine suffering at home. I was asked to give both of you and all your friends the best greetings and wishes of everyone.

The shopkeeper had already been surprised about your last parcel so with this one knows to be much more given than having sent.

About the angels blowing trumpets I will inquire at once by mail. I could have done it personally while staying at Dresden during the past fortnight if I had let the letters being forwarded to me. But in about three weeks I will go on a visit to my parents anyway. I want to help my father a little more because he didn't take any vacations but stayed at Dresden working all the year. There is always such a lot to do and when calculating about a distant date when things will go a little easier something always turns up hindering him to take off for more than a couple of days. I wished I could send him straight to your little rustic cabin in North Carolina so all his connections to his parsonage and Superintendent's duties would be cut off and he would be forced to take a rest. From the pictures you sent me now and some time ago I can fancy the place rather cozy and beautiful and just the right residence to recover from all the rash and haste of civilization and everyday's work. Here in Germany you can't run away to a place like this because the country is too crowded for that, after the war even more than before, because now all the people formerly having been living in the territory now belonging to Poland, Czechoslovakia and Hungary are staying here too. Even the villages are crowded and industries are springing up everywhere.

In the meantime the weather has grown very cold. Today is the third day I have to heat my stove. So you will get a little refreshing air with this letter to your hot Florida. We perhaps should exchange some of the coldness and heat.

When returning from the Baltic Sea this summer I had to pass Berlin on my way home and stayed there for an afternoon. So I could see an American picture I didn't report you about yet. You never told me whether you saw *A Streetcar Named Desire* or some of the other movies I wrote you about, so I think you may not be one visiting cinemas while you may imagine me being crazy about movies. Well, I am certainly not, though I don't share the opinion of most of our Theologists about movies being only destructive. Even Christian pictures are not appreciated of by those who are fighting for orthodoxy and literalism of faith. They call it an outrage and shame that an actor who has played other parts in other plays now acts as a parson or bishop or as a Christian nurse. And they condemn a Christian theater at all, because it appeals with theater-effects to the feeling, the "Empfindung" [sensibility] of people, so it renders only a vague mostly by the emotion going relation between them and men and about a Christian ethical way of life,but disregards the orthodox confession. They think to lead people to Christ only by the way of preaching. But even there and also in the church-music they banish everything appealing to the feeling and call it sentimentality and not being genuine and so appreciate only the "objective" word and music. That means: reading the Bible, explaining it according to the articles of creed and singing songs being just completely otherwise than those melodies people like to hum and sing by themselves. They don't see that the common church-goers aren't interested in their theological calculations and stay away from the services having grown tedious and without connection to the way of living or they go the Methodist or Baptist church where they like the songs of the "English style."

Now I am of opinion that there the fact is forgotten that the heart has to play a big part in religion. God doesn't only want our intellect and will power but also our feelings, and not only the feelings of those who are able to express them in a high musical way and style but also of those who only understand simple songs. And though I know that some music is of more value than the other I don't mix this judgment of art with the reasons to regulate the service unto God and think the Methodists are right in their opinion not to forbid the songs of the "English style." There is no reason why God may not call somebody by the way of his emotion and not his intellect. And it is clearly written in the New Testament that everything we do we shall do with our whole heart. Therefore: to get the whole heart of men directed towards God has to be the aim of the church.

And so also the Christian movies have an important part in the way of mission, because they reach many of those people being never reached by a sermon or Church-paper.

So the movies—not only those showing the religious part of Christianity—might carry something of value towards humanity and are not to be judged upon as being just destructive "a priori" (from the first). On the whole they should have the task to bring humane ideas to people, like a good theater-performance is supposed to do, not just to entertain them as many picture-producers think, offering sensations or nonsense in order to get the money and so poisoning the hearts of people. Real art is not going for the profit but for the ideas it has to carry to people. Church should fight for this and not turn away from movies and theater neglectingly.

So I am always thankful to see a good performance at our opera and playhouse. I could have written about those too, but there was also another reason talking about movies to you when wanting to illustrate thoughts and problems of mine. There are no American plays at our theater any more, just sometimes the performance of a play by Bernhard Shaw or Shakespeare. So I can't talk about something both of us might have seen. And to see a performance at Berlin I have neither time nor money. So you may understand that also from this point it touches me when seeing (or reading) something you might know in the same performance and shape. We are that far apart that this imagination thrills me a little. Especially happy I am when seeing an American picture being of a very constructive kind for this reason, and so it was with the one I saw at Berlin this summer.

Lost Boundaries, a story about a colored medical doctor pretending to be of white race and settling down among white people. Being very skillful he succeeds in acquiring the trust and love of his fellow-citizens. But on occasion of volunteering to the navy his secret is found out about and now all people are furious for being cheated and forget all the good things they had received; only the parson allies with the negro-doctor. And now it was interesting for me to see what influence a parson of a small American town still has because the whole population is assembled there at the Sunday service. And Christian spirit is still able to pervade the way of shaping everyday [life]. In wonderfully touching way the preacher drew away the eyes of his people from color and outside and directs them towards the real value of men, their heart. All this was shown with a fine psychological understanding and an artistic skillfullness. And so without thrilling events and sensations the picture moved everybody following the tale and might have shifted the feelings and ideas of many living in a continent where black and white have to be in constant touch. I don't underestimate the negro-problem of your country for I have read a lot about it and its difficulties, but I was touched by the fine way means of overcoming were shown here and the ideas of brotherhood were not just ordered but carried in such way to a population whose country was built and defended with the sweat and blood of white and colored men, having the same faith unto the same Lord and Saviour besides.

The manuscript I had revised isn't printed yet because it is not easy to attain an allowance for paper, though the license for printing we have got. But I think in some weeks the printing may be started. I would be very interested in the story of Franklin's life by your aunt and am sorry printed matters aren't allowed to enter the DDR. If it is of a political tendency I might *(not?)* get it passing the censorship. The book of the nature-poems still stands on my shelf and I often take it down and read one of them.

With all the best wishes and loving greetings, I remain

Sincerely yours

Christoph

Just in time for the holidays, restrictions are back and severe. Christoph is trying to work around them by having packages go through West Berlin.

November 30, 1952

Dear Miss Grover,

Dear Mrs. Blakely,

Until the pulling-out of the train carrying me away from a one-day visit at Berlin, there is half an hour left. And I long to spend this time by sending you some greetings. It was a nice day, but filled with too many things, so no leisure was possible. I visited an acquaintance of mine in the American Sector of Berlin. He offered me to take care of letters and parcels, and send the parcels to me, because custom-restrictions are now that severe, that parcels from foreign countries are charged too high. But to the American Sector they easily can be shipped and from there I may get them at a visit of mine or it will be forwarded to me. I want to write you this, because you certainly will have heard about those restrictions. Regulations change often and so it might be possible, also, this restriction will be lifted again. So we are able now to go on sending small packages. We will try to get off the next soon. Material conditions at home are now lowered a little, and quite some things are short again. But please, do write about this and about parcels only to the address at Berlin.

My thoughts are often with you, and I suffer a lot from the rent which is going to cut deeper and deeper the world into two parts. May the Lord spare us the most terrible catastrophe, a new war. All my good wishes for Christmas go with this letter to you and all your friends.

Sincerely yours,

Christoph

Restrictions are becoming more severe, and evidently mail is being read by censors.

February 12, 1953

Dear Miss Grover, dear Mrs. Blakely:

With a sore heart I am writing this letter. Deeper and deeper the rent is growing between the two parts of the world trying to blow up all the bridges connecting them. Restrictions and atmospheres are created to keep off people to stay combined. You know how happy I was to have won the friendship of yours and what strength I drew from this bond between us, and so you will understand that often I have to fight the tears when looking upon the way politics are developing and when seeing you sliding farther and farther away from me instead of my constant hope that the love between people of different countries will overcome the hate and ward off a division of the world. But whatever happens I will keep up the bridge between us and if it might at last only consist of my constant thoughts about you and the remembrance of all the beautiful and wonderful hours rooted in the meeting of our hearts. And the bridge going from my prayers through heaven to you will only cease with the cessation of the beating of my heart.

For the time being it still is possible to correspond. You also may write to Leipzig and receive mail from there, but nothing about politics, economic problems or Berlin, etc.

Parcels are allowed too, though only half a pound of coffee (ground or instant) and half a pound of cocoa (including all items made of chocolate) will be forwarded. If there is more, the whole item gets confiscated. So it happened with the parcel sent to Dresden. Coffee and cocoa was confiscated, all the other things delivered. At Leipzig I asked to return the parcel. I hope you will get it. You may—if you like—pack it according to the restrictions and send it to Dresden and to Leipsig. Also there may be packages after a distance of a couple of weeks. Tea is not restricted, only coffee and cocoa. There also is a way of getting packages from Western Germany. There, *one* pound of ground coffee (no instant!) is free to be sent from America each month, but only if there are two pounds of normal food added. My father has a friend in Western Germany who will forward things to us. His address is: Haufe c/o Pfarrer Muller, Selb G. Hof, Bayern, American Zone, Germany.

We also sent you two packages with figurines (a car loaded with furniture) and are anxious to hear weather they arrive. This you might write to Leipsig because [it] will be some weeks until I revisit Berlin.

My best wishes and my longing is going with this letter.

As we see in this letter, restrictions are getting tighter and quantities in excess of what is allowed are being confiscated, as is printed matter. Stalin died in March 1953, and the uncertainty of what is to come is affecting life for everyday East Germans.

Christoph's description of the paper he is translating is interesting. It appears the American professor who wrote the paper is objecting to public schools not having enough religion in their curriculum. This was during the McCarthy era in the U.S. "Under God" would be inserted into the Pledge of Allegiance the next year. It's surprising to me the paper is slated to appear in an East German publication.

April 8, 1953

Dear Mrs. Blakely,

Dear Miss Grover,

I was very happy to receive the letter announcing the safe arrival of the little package at the home of Miss Grover's family. In the meantime, you will have seen its contents and I hope the first greetings of this kind sent to you this year will give you joy and hope that the bridge of our friendship also this year will be used successfully. I was so glad to hear the news about the safe arrival! I succeeded at once to get some more Erzgebirgian figurines when staying at home over the Easter-holidays, my mother will pack them and mail them this week.

Concerning the Christmas-parcels you sent at the end of last year I sorrily have to inform you that it happened as I wrote you before. The parcel for my parents was opened at the custom-office at Dresden and the coffee and cocoa were confiscated, the other items were delivered. At Leipzig I declined acceptance of the parcel intended for me and I succeeded in sending it back. But now I am worried and sorry to have put upon you the trouble about this parcel being shipped back to you. Though it already means a very sad thought to see such precious things (that very needed and longingly awaited for) after their safe shipment over the ocean going the way back they came, the most troubling feeling is given me by the fact that now you will have even to pay additional postage for that useless circular tour, unpack it and split it into parts according to the regulations and reship it again. What trouble and extra-work laid upon your shoulders! I certainly don't know how to make up for it.

But even another—not necessary—misfortune happened. Mrs. Blakely announced some weeks ago a package with some ties. In my phantasy I already imagined them to be tied around my neck and constantly reminding me of you and everybody—asking me about them—telling about you. Now this package was addressed to Dresden and it arrived just at the time my parents were on a few days of vacations. They were at home all the time of last year because on account of the stress of work my father couldn't take any rest last year and was persuaded—because he badly needed some rest—to take some days off in February of this year. And just during these days the package arrived. A friend of our family, keeping the house, didn't know what to do, when he got the card from the custom-house, but because he had heard about the misfortune having happened with the parcel of Miss Grover he thought the contents of this package also were consisting of same coffee and cocoa and because a decline had to be told to the mailman immediately (before knowing about the contents) he was asked to send the package back. Can you imagine my disappointment when I got the news about that! Now I only can hope the package will be returned safely and Mrs. Blakely won't be too angry about the postage. It wasn't necessary to have had it to be returned. Because only coffee and chocolate-items are restricted, of which half a pound of each can be included (a German pound is 500 grams). Now I don't know what to say; whether I can dare to ask to return the ties another time, and—if there was coffee in the parcel—to reduce it to half a pound (250 grams)? Daring it I have an uneasy feeling.

You ask me about custom-duties, whether they are high and to be paid for each parcel. Well each parcel comes to the custom-house and the addressee gets a card informing him to fetch it. There it has to be opened by the addressee and checked by officials. The main reason for it is to make sure no weapons or antidemocratic propaganda is shipped in. All printed matter will be taken out therefore. Then each item, all the cans and packages are checked for their weight. The charge put upon them is very low and gladly paid. So I paid for the tea you sent last fall only 62 Pfennigs, somewhat over half of the price one has to pay for an haircut.

Yesterday I returned from home where I spent the Easter-season. We passed wonderful days of family life and my mother had to set all her skill to work to satisfy all the guests. The first sunny spring-days were enveloping us and spending their share to enlighten our hearts. The day before yesterday when opening the door of our shed too near-born goat-kittens were trying to spring around their mother and catch her nipples. So the spring had even pervaded the walls of our shed, and the two small lovely creatures and all the blossoms surrounding our home evoked the hope and confidence the cold days of the world-affairs also might pass and the spring for a lasting peace between the different nations may be greeted.

At present I am reading the *Forsyte Saga* by Galseworthy, and though it is a volume of over thousand pages I enjoy it very much. I had started to read it about two years ago and put it aside again after going through some pages because I was too much wearied by the many names of family members and their complicated relationships. But after having listened to a Radio play about the Forsyte-woman Irene I was so much interested to start it anew and this time I am enthusiastic about it. Nevertheless, it strikes me how far we have moved from this world of the Victorian age, this life among the members of the upper middle-class, how lavishly they lived and how much they always were concentrated upon property—relating the value of everything to it. Well, people of the last kind we might meet nowadays too!

You ask about things you might put in as change for the restricted coffee and chocolate items, and that reminded me of a trouble I always experience when baking a cake by myself. I think I wrote you already years ago that now and then I like to bake a cake by myself in a little electric stove. So far the output was mostly quite tasty—though not as good as my mother's cakes, because the ones made by myself are of much simpler way of preparing—but the artificial flavour (lemon, vanilla etc.) is not of the best effect. So I would like to ask you to help me in baking. I wonder whether you have real vanilla.

For shining my shoes, I am in lack of a light-brown and black cream and a mouse-gray powder.

Now I feel like a child—being not moderate at all—when writing down his wishes for Christmas.

A fortnight ago my knowledges of the English language were very useful for some professional work. A theological essay of 8 typed sheets was sent from an American professor of theology to our seminary in order to be printed in the *German Theological Journal,* so I had to translate it. It was about education and moral, and I was left in uncertainty whether "morals" can be (as required by the contents of the essay) translated with "Weltanschauung" (conception of the world, philosophy of life). The essay tends to show that the way schools are established and taught in always reflects the ideas and morals of the older generation which wants to inherit them to its children. After a survey of different European countries and their differences in relationship between church and school the author expresses that schools in America being in the hands and under the influence of some small antireligious scientists are under and antireligious tendency. But since the population doesn't want her children being education without a feeling for the importance of religion, soon the public state schools will be a mere shell while children are sent to parochial schools, if there will be no change in the education-policy. I wonder whether the judgment of the author covers the real facts.

With the best wishes for your welfare, I remain yours truly,

Christoph

Please give my greetings to all who remember me.

Greetings also from my parents and brothers!

April 14, 1953

Dear Miss Grover,

Today I received your parcel. The joy about the safe arrival and delivery was of special joy to me since after the long period during which the bridge between us didn't function according to our longing, it was the first parcel after the custom-restrictions which was handed over to me as being completely OK. The big and small jar of instant coffee were a little under the allowed amount and because there can be prepared quite a number of cups by them it means a big help for my work, so I want to inform you at once about this safe arrival and my happiness to receive this gift of your helping and friendship. Many, many thanks to you and also to Mrs. Blakely, who also will be glad that after the mischief of the last parcel now I can ascertain you that delivery without loss is possible again.

The water I put upon the stove is just boiling, so I will prepare the first cup and drink with special thankful remembrance of yours.

Sincerely yours,

Christoph

Christoph is again in Berlin and comments on how easy it is to go from East Berlin to West Berlin. Alas, prices keep him from buying much, and American novels are not allowed in East Germany. He enjoys An American in Paris. *Note travel to Berlin is becoming more difficult.*

May 1, 1953

Dear Miss Grover,

Dear Mrs. Blakely,

Another trip to Berlin I won't let pass without having

sent you my love from that space of the part of the world, which is connected with your country but nevertheless being possible to enter for us living in the DDR. But it is getting more and more difficult to go to Berlin, and each time I don't know whether it is the last time. They ask in the trains to Berlin what business we undertake this trip for, and they always are suspicious to face an agent, a spy, a black-marketeering guy of a refugee. Being at Berlin it is comparatively easy to cross the border, because the subways are running through all sector of the city yet. And it always renders a strange feeling to leave a subway-station of Eastern Berlin and see at the next one different slogans, different newspapers, see chocolate, lemons and other things, and be in the western world. It's strange to meet now—as some minutes ago the Russians—English, French and American soldiers on the streets and to be able now to render American movies (if we have the wester money, needed for entrance. The exchange rate is nearly 1:6. So one lift with a streetcar costs me the equivalent of 30 rolls of white bread.)

Yesterday I saw an American picture and I always imagined you *might* have seen the same. I only felt sorry because it was synchronized because I would have lived to hear some American. The German title was *An American at Paris,* but I don't know whether they translated it exactly. It was a revue-movie but of a rather good and skillful performance. Besides I saw an Italian picture.

So Berlin is a very interesting city giving you a view of the whole world. With sore heart I have to pass the bookshops where they sell American novels. Even if I had the money to buy some, I couldn't take them with me, because I might get checked on my way back and no wester printed matter is allowed to be imported. Nothing is allowed to buy in western Berlin, anyway. In the zone, it is difficult to get new American literature. Last I read *Cry, My Beloved Country* by Paton, but I only succeeded to get a German translation. Nevertheless, I was so touched by this book that I often had to stop resting and look through the window into some blue spring sky because my eyes for veiled. A colleague of mine told me he knew somebody who owns *Of Glory and Power* by Graham Greene and I am anxious now to get it. We all hope the present strain of political disagreements will be settled and solved peacefully soon, so there will be normal connections again, not least the cultural connection and relation between our peoples.

Your parcel arrived safely at my parents' home. So both gift-consignments of yours are now used by us to overcome the nourishment-difficulties that have arisen since this winter anew. Many, many thanks for your love! Everybody was delighted about the Maxwell instant coffee. They thought it the best; but it might be only the newness of this kind. My father asked me to thank you specially! You know, he has to drink coffee for his health.

I feel very near you; as if I could reach out my hand and touch you. Perhaps already the day after tomorrow you will read this letter. The Lord bless you and keep our friendship under his protection.

Sincerely yours,

Christoph

As I mentioned in the history section, there was a massive general strike in June 1953 that saw nearly a million East Germans, including 100,000 in Berlin and 40,000 in Leipzig participate. The Soviets showed up with tanks to quell the uprising. This is what Christoph is referring to when he mentions the "disturbances" Ms. Blakely and Ms. Grover had heard of. We have to imagine Christoph has to watch what he says and certainly can't elaborate about what happened.

July 2, 1953-

Dear Miss Grover,

Dear Mrs. Blakely,

I was so happy to receive your wonderful precious parcel. Also the one you send to Dresden was received by my parents. Both parcels were delivered without difficulties and caused immense joy among all of us. The extract for baking is wonderful, never a self-made cake did taste that delicious, and nowhere a cake of that quality can be bought. I am really proud when serving a visitor a slice of it and I wished you could be my visitors over next Sunday. It would be a wonderful weekend and I would make the best picture of you with my photograph. Since quite some years me and my brother John have the hobby of photographing and have a nice little box. The quality of the pictures depends on the love they are taken with, naturally one can't shoot right away as most amateurs do but has to know some tricks and the handling of a box. Besides we have in Jena-city (DDR) a very good optic produced which famous all over the world, most of it is exported, but some sold in the DDR. Today I include a picture taken by my father after he said, Now, we will have one of you smiling! and he tried to tell me some jolly things, about which attempt I already had to smile.

My shoes are now shining as never before, I really start to like polishing them. The mice-grey shoes are suede ones of my mother.

But most of all—as ever—we rejoiced about the coffee. My father always asks me to tell you how much it helps him, since he needs it for his health and cannot buy it; he said he wouldn't know what to do without you. But also for my work it is of great help. A cup of coffee, quickly prepared, and what didn't want to go goes on now. And I have a lot to do. This summer I want to finish my dissertation, so I can deliver the writing work this year and pass the Doctor-examination (the oral one) next year. I am writing all days now, and will only stop for a fortnight-vacation. Besides I again overtook some services for minister who want and need vacation but have no substitute. For all this you help me successfully do it. So we all thank you very much!

I am troubled about Mrs. Blakely. It is a long time since I heard last of her. I also don't know whether the ties were returned to her and whether I may wait for them. Please, do write me about how she is feeling.

About the disturbances you might have heard about you need not trouble. All is quiet and normal again. Certainly you also will have heard, that our government had corrected some mistakes and so some improvements are awaited. Also the boys and girls, which were expelled from school on account of their belonging to the Christian youth organization are taken back, and some who were imprisoned have been released. So we all are glad and thankful. There are no new difficulties of import, though the restrictions aren't lifted yet as I hope will happen too. So all is going on unchanged.

I remain as ever with all good wishes your truly,

Christoph

Christoph expresses his views on various approaches to religion and baptism. He also notes many positive changes that came from the general strike.

August 30, 1953

Dear Mrs. Blakely,

Dear Miss Grover,

I am getting more and more worried, and I pray to God

there may not be a reason for it. Many weeks passed by without hearing from you. The last news about you, dear Mrs. Blakely, I received 5 months ago were not very reassuring and since that time I didn't hear anything from you anymore. I wished I had a private plane to come over and look after you myself. You, my dear friends, have become that dear to me during all the by-gone years that my feelings rather consider you as belonging to my family. Your pictures are on the walls and there goes no day without mentioning you and enclosing you in our evening-prayers. My greatest hope was to see you once in this life at least, and I even consider the latest political developments under the aspect that they may bring up the chance to go from a reunited Germany on board a ship and pay you a visit. Perhaps you by yourself may wonder about all this, but you shall know that the Lord selected you to keep me going on studying Theology. When during the years after the war I noticed that not even by the means of shell and bombs people turned inside and kept the light of love still shining when all the fear and dread, all the promises given during the terrible days and nights the death had be a raving king were forgotten there were hours of doubt and despair; also when the girl I loved turned away from me. It was the picture of my father that kept me going, but not always with the joy it should have been with. And then there was your friendship and love going on and going on, never relaxing, and I felt ashamed. I learnt, it never is astounding and most important that most people—even when talking about church and Christ—think just about themselves, but rather the most important is on the contrary that there were some who are not. I met some of them, not many, here and there, mostly unnoticed by the big world of big words, but there are a few. And seeing this I was glad not to have changed to the multitude. I learnt there were always some who carry the spirit of Christ onward to the next generation, and it is the task of those belonging to the young one to win some again, to carry onward too; not many —my belief on this faded away. There were too many who acted one day as if they were won and yet had forgotten everything after a week, and there are so many who call themselves to be Christians and are worse than the Atheists. Those Christians nearly had overrun my faith before I learned what I know now, that Jesus picks his disciples here and there, but they are few, and many who call him Lord, Lord he won't recognize. But everywhere a Samaritan is going through the world and stopping at places his help is needed there is Christ's spirit, and not always in those big churches where so wonderful big words are spoken, but outside—the minister quarrels with his wife and his congregation-members go where they have the best profit. It may sound paradoxically, but it was the church-people who nearly brought me to giving up to become a minister. I couldn't believe that Christ had died in order that people should become just religious, being saved by a simple acknowledgement of divine facts and dogma.

Next Sunday I have to represent a minister so he can take his badly needed furlough. His church is a very small but homely looking old country-church in a suburb of Leipzig. It was left from a time this suburb had been still a village. The houses formerly surrounding it were replaced by bigger ones, fitting for a city. So you stroll through the suburb, the streetcars running by, and suddenly stay in front of this village-church, much lower than all the houses nearby.

About a wonderful word I will preach. "Do not look at his appearance or the height of his stature, since I have rejected him, for the Lord does not see as man sees, for man looks on the outward appearance, but the Lord looks at the heart" (1 Sam. 16:7). A word likewise frightening as consoling. What puts the men at rest in their relations to each other has to disquiet them in their relation to God. I wonder what the world would be like if people could look into the hearts of their fellowmen! Everywhere egoism and lusts! Nobody then could cheat the other anymore, and many a girl wouldn't have married a man whose wife she had become, and vice versa. But as it is, one judges upon the character of the other one from his outside appearance, his beautiful sight and his beautiful words, and forgets—as if we shouldn't know! —that nowhere else greater mistakes can be done. With the Lord it is quite otherwise. We can't cheat him, he doesn't believe our pious phrases and our clever constructed arguments and excuses, why we for instance do this and act that way, though we should act a little different. We think he believes our arguments, when we—even —explain to him why we couldn't act as a Samaritan at this time the opportunity arose. Also the Priest and the Levite had good theological reasons for passing; according to the Jewish religion a man beaten down was thought upon as [having been struck down] by God; should they intervene with the punishment of God? Or: Somebody who touched a dead one (and perhaps the wounded one may have died under their hands) is impure and unable to render the service at the temple; and they went just there for it. The service at the temple was—and also today may be—more important to them than helping the one needing them. How hard the judgment of Ged will be, because he can see quite through us and sees the real motives behind the theological arguments! We can't stand upright with our backs in front of him (even not with the confession that we can't stand upright in front of him; because it mostly is a confession of lips) as we may stay upright within our community on account of the outward behaviour having been always with the borders of the law.

But about this word there can be said just the contrary too! So glad people are that nobody can look into them (if they had a window in their fore-head, that would be a running for milk-glass!) and so worried they must be about God looking at their hearts, so sad—on the other side—they must feel, when not being able to open their soul to another one completely, in vain striving for words, and when they also feel never to be recognized by the other one without misunderstanding, but so happy then can be that the Lord even understands all what they are unable to express. The longing, somebody to have quite near to their heart and understand its beating, never quite fulfilled within the relations between men, becomes true in in the relation with God. The loneliness of heart, a feeling never completely overcome even in the relationship of a loving couple, here with God it is overcome. We can have the feeling of nearness, understanding and security. How often people cry (when their actions had this or that effect for which they are blamed): We didn't intend it, we didn't mean it, we wanted only the best! but they are judged upon their deeds. With God it is different; he does not just judge upon the actions but also understands a heart which was willing to do the best but had been overcome by weakness or circumstances or lack of knowledge. So hard His judgment is, because He never is content with outward appearance and lip-service, so gracious it is for the same reason; he understands those who fail but always are willing to try and start struggling and striving anew and suffer under their impotence to be as they long to be.

Well, now I even had meditated about the sermon of next Sunday! But as you will understand, my thoughts were occupied by this word of the Old Testament and I put them down. To me this word often meant a great [consolation]. It consoled me more than all the sayings about: if you repent, you will be forgiven and if you believe you will be saved. I never knew whether I repent or believe in the right way. The way the churches preaches it often seems me too simple and dogmatical. I cannot imagine that repentance and faith only mean a mere nodding of the head towards the truth. And many things the church says one has to believe in, in order to be saved I cannot believe in. I cannot believe that God Father and Son are two persons and one God; and I cannot believe that the holy spirit is an improving power of some magical kind given through the water of baptism, so that—according to the Lutherans—baptised child has a better quality than another one. Concerning the last point I am more of the opinion of the Baptists. Baptism is an act by which men document that they want to be followers of Christ and by which is given know to them that so they will be accepted by God. I often have hard discussions and many Theologists and men of church consider me a bad Lutheran. But the knowledge that God is looking at the heart keeps me upright; he knows that I am looking for him everywhere. Often I think it is more important to try to do what he *says* than to speculate and believe in certain destinations and definitions about what he *is*; I never found him, when trying to imagine his being, but I felt his nearness, when I met a Samaritan.

With these thoughts of mine, often occupying me when the wind is moving the trees outside my window and the clouds are passing over them to the horizon, a part of myself goes over to you, they may confer to you more about myself than a lot of words written about outward things.

During the last months many things happened giving a new hope to the world. Changes took place in my country which are of great importance. There was issued a law that all who had fled from the DDR may return, no harm will be done to them and they will receive all their property, the farmers their farm and the shopkeepers their shop, etc.; and it really is done. Also with the church an agreement is reached the Christian youth is not persecuted any more: those who were expelled from school are allowed to return, and those who were imprisoned are released. Financial means—formerly given to the heavy industry and police—are now promised to be spent for producing or importing food and cloth and shoes, so there shall be real leather shoes for all people. The reparations—a big amount of industrial and other products—we had to pay to Russia will be cancelled with the end of this year, and big factories, confiscated by Russia after the war, will be given back. If this all will be realized we have reason for a better life. But more than about the material improvement we think about the hope, that all this will lead to an understanding between the governments. And when the border—running through Germany and dividing it into two parts—will be cancelled and Germany a neutral state, then the way is free also to pay you a visit. Please, pray with me that the policy of this world really will lead to peace.

About parcel delivery there are no changes. Parcels still are delivered without difficulties. I write you this, so you may not mix the news about the distribution of parcels at Berlin, only those are forbidden on account of their propagandic tendency. But all private mail is going on unhindered. About the safe arrival of your last two parcels, I already thank you via airmail, but I want to do it again, because we were so happy about them, and the coffee meant a big help for my father and me.

Now I am looking out for a letter telling me all is O.K. with you. Please, don't forget your friend in Germany, as I never will forget you.

Sincerely yours

Christoph

Christoph describes the subject of his dissertation. He will complete it. I have seen it referenced as "Die sittliche Rechtfertigungslehre des Paulus" ("The Moral Justification Doctrine of Paul") (Halle: Max Niemeyer, 1957).

September 10, 1953

Dear Miss Grover,

A few days after mailing an ordinary letter to you telling you I was worrying about you and Mrs. Blakely, I received your letter and was much relieved to hear about you being well. I am that glad about it that I at once will write an airmail to tell you about it; I hope it will overtake the other one, just flying over the ship carrying it. I want to wish you all good luck for the start of this new school-term. The fall-term of our Seminary starts next week, so both of us soon will run together again in a similar work. My dissertation is going on steadily, though not as quickly as I hoped. I write about the relation of mercy and reward according to the Pauline letters. It inquires about the problem whether faith alone will be sufficient for salvation or some moral conditions will also have their influence upon the judgement of the Lord; if the latter is right—and I think it is—then those Theologists are wrong who claim that God is doing everything and men are depending absolutely on grace. There are with Paul, a lot of admonitions towards Christians, and they are combined with reward and punishment. This would be superfluous, if the only condition—as Lutherans claim—would be to accept grace which then would do everything. If you are interested in it, I will tell you my view how to solve the problem of faith and merit.

Of the oral examination I am not afraid, as long as I am not asked things I don't know. And about everything one cannot have all the knowledges; my memory sometimes leaves me in a lurch, and I often have to reconstruct or even invent.

Dorothe is the greatest joy to my father, her granddaddy; she starts to talk, and always when she hears a church bell ringing she looks intensely and exclaims, "Opa!!" (Grandfather) Her parents have a lot of work. Both are very industrious, but since his products are rather cheap but all he has to buy is expensive my brother won't get to something.

From Mrs. Blakely I didn't get any news, and after hearing from you the same I am getting even more troubled. If you hear from her, please do drop me a note.

About the election in Western Germany many Americans will rejoice; but for us it means disappointment. Germany will stay divided, because Russia will only consent to a neutral united Germany. And it could be otherwise! After German troops had invaded her country that far!

Sincerely yours always

Chris

Christoph is gearing up for a busy semester.

October 24, 1953

Dear Miss Grover,

Drinking a cup of wonderful coffee to give me the energy to go on preparing my lecture I have to remember you and put aside my work in order to send you a few lines of thank for your precious parcels which have arrived safely already a fortnight ago. I waited until I got news from home that also the box you sent to my parents was received happily. As always, the joy was great and my father asked me not to forget to give you his thanks too. The coffee means medicine to him and so he feels himself very much dependent from you. In the meantime, also a letter from Mrs. Blakely was received and we all are relieved to hear she feels well. At present, I have a lot to do. Most of all I have to work for the lecture I deliver this term. The theme sounds very easy, though. "Introduction into the New Testament." But since a survey about all the problems has to be given it means for me a sound preparation. Not only all the circumstances of historical kinds an epistle was written in have to be talked about (time and place of the congregation it was sent to, the ideas of the enemies of Paul he fights against, the question of a letter being genuine or pseudo, etc.) but also all the theological questions treated by it have to be summoned up. The difficulty arising concerns the method to deliver such a heap of stuff in a way being understandable to students. Presently I examine how to explain the 4 parties in Korinth and find their ideas in the letter to the Corinthians (1 Cor. 1:12). At Christmas I promised my father to overtake the sermons at home, so I will be all the holidays with my parents at Dresden. All good wishes!

Sincerely yours

Christoph

Christoph waxes philosophical about Christmas and is putting the finishing touches on his 800-page dissertation.

December 19, 1953

Dear Mrs. Blakely,

Dear Miss Grover,

I hope this letter will reach you just at Christmas-Eve and thus bring you all my wishes of blessing when you are sitting under the lights of the Christmas-tree. Also, my thoughts will go to you when in the round of my family at home we rejoice about the birth of the child in the creche who wants to bring this world the peace with God and therefore calls it to brotherly love and peace. How could we forget to remember you at this day of the greatest love the world ever saw.

Many parsons in their Christmas-sermons will blame most of the people to have changed the day of heavenly love into a festival of mere human love, because the central point is forgotten and all thoughts are concentrated on the love between them showing itself in gifts and by the hope to establish peace. I think nowhere Christmas is more celebrated this way than here. But I think we shouldn't blame them doing wrong but rather ask them why they just at Christmas have the feeling of love and peace in a special intensity, why just at Christmas those who quarrel all the year-long for reconciliation. In all the so-called "secularization" of Christmas there seems to me still to be the hidden well of the real Christmas. And all the feeling awakening even in the hearts of the Atheists when they listen to their children for something they have lost. But all those feelings, also those of love, pass with the passing of Christmas and everyday' life buries them because there was nobody who—instead of blaming them for their secularized "sentimentality"—did show them the well all was coming from and so give them the possibility to keep the light of Christmas shining through all the year. If they only knew what it was the Child of Bethlehem who knocked at the door of their hearts they perhaps might willingly have opened it on account of all those feelings they so suddenly experienced at Christmas, and then perhaps they might have followed the child all his way healing the sick and consoling the lost and longing ones and showing them the love of God by the happenings of Easter. Then the well of love would flow not only in the present short time of Christmas but always. I am so glad when meeting people in whose heart Christ really was born—not in the dogmatical way of the church, but as an inward light and strength. And so just at Christmas surrounded by all this rapidly passing "sentimentality" I am urged to thank God he let me know you. So many years we know each other, letters are crossing the ocean and signs of love following them.

Your Christmas gift was not only a mere Christmas-gift, but one gift in a long row of acts from your helping hand. And that made it special dear to me, because it was a real

Christmas-gift and not an exception at Christmas-time. With my thanks to God, I remember the words Paul has written to the Corinthians, 2 Cor. 9:8-15.

My vacations already have started. They are four weeks this time. But with the exception of the holidays, I need all the days for the final conclusion of my dissertation-work. In January I hope to put the last point. Then I am in trouble to find a good typist to dictate the about 800 pages. Wouldn't you like to do it? Or you might send me one of your pupils! If there wouldn't be the German language and the far distance!

During the holidays I have to deliver two sermons at home. At home I also will take some pictures this time so you will have some new snaps of the family. The "Introduction"-lecture I have to deliver will be continued next term, because it goes for a year. Beside that I have no other lesson as the professors deliver them; it is my first lesson of this kind. The other work is not that difficult: reading the Greek New Testament, the Hebrew and Aramaistic Vischna (Talmud) of the Jews and sources (Greek and Latin) of Hellenistic or Orientalist religions.

I enjoy my work, but it keeps me going. Your coffee means a big help. Both of your parcels for me and my parents I have received and the air-mailed letter of Nov. 2 needed 5 days to reach me. Many thanks for all your love. I hope the Christmas-package I sent off a while ago will arrive with this letter and so be on your Christmas-table. After the holiday-rush I will send you a new surprise. All my wishes go with this letter, wishes for Christmas and for the New Year. The Lord may protect our bodies and our souls and our friendship.

Sincerely yours,

Chris

Christoph is suffering from eyestrain, which is not surprising. He is excited about a church conference to be held in Leipzig that will include people from both Germanies. Travel restrictions have been lifted, probably as a result of the general strike in 1953, but note currency cannot be taken out of East Germany.

May 12, 1954

Dear Miss Grover,

Dear Mrs. Blakely,

Weeks passed without writing you how glad I was to receive your letter of February 28. And also, I tarried too long to thank you for the very precious gift-parcels you sent again. There were two to Leipzig and three to my parents in the meantime; they safely arrived and are of such big help to us; my father is enabled by them to go without headaches and also for my work the coffee is of invaluable assistance. You really do a part of my work by your love.

The reason for me not writing during the past weeks was a weakness of my eyes. I had overworked them by too much reading and writing, and so I was restricted to only some hours doing this kind of work. I had to use those hours for preparation for my lessons and for dictating my dissertation. I had calculated to have it finished by now, but just 445 pages are written. Anyway, it is more than half of it and I certainly will finish until summer. Then times will be easier for me and you will get a long letter, for there is so much I long to tell you. And then I also will have the time again to read English books. For lack of training, I fear I have got rather worse with my English style. The oral examination has to be postponed until late fall, since all the professors have to read the dissertation first. At present I feel quite some improvement and I can prolong the time of reading and writing. So I don't want to wait any longer to write you. If I had found somebody I could dictate English to I would have done so with a letter to you, as I did with some of the necessary German correspondence.

In July there will be a meeting of the German Evangelical church here in Leipzig. The whole city will be overcrowded by people from all the parts of Germany assembling in churches and halls to listen to outstanding speakers or to discuss various problems in smaller groups. For some days the life of this city will be colored and stamped by this "Kirchentag," and to all the non-Christian people will be demonstration that Christianity is not a dying ideology but a strength living in people old *and* young. Many thousands of Christians also will come from the western zone of Germany. There is no hindrance any more for crossing the border dividing Germany. I formerly wrote you about the difficulties which had to be overcome if somebody wanted to visit his relatives living in the other part of Germany. There were times it nearly was impossible. Since last year the restrictions are lifted. There are still controls at the border but everybody can cross it. Only the different currency makes it impossible for those who have nobody in the other part to pay for them, because there is not allowed to take any money along. But at the "Kirchentag" this difficulty will be overcome. So everybody will make the trip to Leipzig who wants to demonstrate by his presence the unity of the church of eastern and western Germany. Well, and I may add by experience, *we* demonstrate by our friendship the unity of Christianity of the whole world.

I will give you a report of the "Kirchentag" and I will take some snaps to send you, so you also will get an impression. And here we are at the snapshots. The pictures you sent me are very nice, they really look like made by a professional. And most of all, now I can imagine in which surroundings you do your living and working. I didn't know you had a flashing bulb. By it you are able to take wonderful shuts by surprise. I often wanted to buy such flashing bulb myself but I haven't the money yet. I take my indoor-pictures with the help of three big bulbs; this even has the advantage to let me work with grades of shadow. It's my hobby to photograph, and to develop and copy the pictures myself. Mostly I like to take snaps of children. I enclose some of little Dorothee; also some pictures of my parents will go with the ordinary-mailed letter. I hope they will close the contact between us. We all would be glad to have you as a guest among us some day. Perhaps this hope is not in vain, if at last normal peaceful relations between the countries will be established. But in the meantime, our thoughts travel to you and the pictures of you put up on our desks remind us constantly of our dear family-members in America. All our wishes for your health and happiness go with this letter. As ever sincerely and thankfully yours, Chris

Germany is having a cool summer. One summer I visited friends who moved there from Romania after the Romanian revolution. They apologized for the cool weather. I told them I was enjoying it, since that time of the year in America we had to pay for cool air.

Christoph gives a fascinating description of the religious conference, which was very well attended. He writes people slept on floors to provide those who visited Leipzig for the event with places to stay.

His dissertation is being reviewed, and he feels good about it.

July 30, 1954

My dear friends,

In spite of the wet-cold rainstormy weather outdoors it is a beautiful day since I have to answer letters of both of you. I am so glad you are well and have visited me once more via air-mail. When I go home tomorrow starting my vacations there will be quite a family-talk following my translation of your letters. When I read about the continuous heat in Florida we all will agree that a little bit of weather exchange would do good to all of us. What you have too much of heat we have too much of coldness and rain. There was no summer at all until now. If your letters had been living persons they would certainly have [caught] a cold when entering Europe. Writing your letters it will have been a toil on account of the heat, and when delivered here it was a difficulty for the mail-woman to discard them and read the address without getting all blotted out by the rain, and when read by me—6 days after you mailed them—I was sitting beside the stove. There were days in June and July we had to heat our rooms; and we always went with pullovers and raincoats. The letters—if they had the ability of reflecting thoughts—would think they were sent from Middle Africa to North-Scandinavia. A fortnight ago we even had a flood at Leipzig. Some streets were covered with water nearly up to the first story and people had to use boats for going out or coming home. Some parts of the city resembled Venedig [Venice]. At my home the water stood just at the door when the flood started to fall again. Now everything is normalized, the cellars that had been filled with water were pumped empty and dried. But the weather is staying like April. And I intended to go to the Baltic with my brother John on August 5th! If the weather doesn't change there will be no swimming. Couldn't you send some heat over? We gladly would leave you some of our coolness and rain. People talk about Atom bombs causing the crazy weather, but I haven't heard any scientific explanation giving this connection.

My dissertation-work is finished. Professor Leipoldt already has read half of it and is quite pleased, so I certainly won't get it returned for a polishing up.

My eyes have completely recovered. Perhaps it wasn't at all the reading and writing causing the weakness but rather a wrong nutrition. Because I am living for myself I am often not in the mood to cook something or to board a streetcar in order to go into the city for a restaurant, and so a slice of bread and an egg must do it. Each time I come home my mother is blaming me for living what she calls: without regular order; one should have regular times for dinner and not only eat when hungry, and besides one should have at least one real dinner a day. So since some time she more and more hints how nice it is to be married. Just the contrary to her point of view years ago when she warned me not to decide too quickly. Well, I am not against marriage, but I don't fall in love anymore, and I think a little love should belong to a married couple. Years ago my heart was more vulnerable, but today I am perhaps too skeptical. Sometimes I wished I would meet a real Christian girl I could lose my head when encountering. But the days I wrote poetry (I still past have a big collection of it) have [passed]. At that time I studied Anglistic. And I also took a course of American History. I was reminded of it, when reading about your plans to teach such course. How I would like to join and have this repetition; perhaps I again would sit beside an eager student. But I will join your course with my thoughts. Where do you start? What time will you talk about the French war and what time about the Boston Tea Party and the Louisiana Purchase? Do you have a time-table of your work? If you had I would exhume my history book and my notes and repeat according to your time-table. So I would be an invisible pupil of yours and your class always would be for one pupil bigger than there are present.

From 7th to 11th of July here at Leipzig the big church-meeting was a success in spite of the rain. 60,000 Christians came from all over Germany to take part. And over the week-end there were even 600,000 who were assembled at the big

Rosental-Meadow. My father and my brother John—my guests at the church-meeting—were standing with me in the middle of those 600,000. I let myself be heaved up on the shoulders of my brother and a theological student and took pictures round in all directions. I enclose them, so you may have an impression too. (1) Through three hours people were standing thronged and listened to the prayers and addresses of the outstanding church-leaders. A while high cross put up in front of a big gallery filled with hundreds of trumpeters was concentrating the eyes of all. Bells were put up to ring at the beginnings and the end. Loudspeakers were dispersed over the whole area, so everybody could understand the words of the speakers. The big place surrounded by white flags with a big violet cross was filled by Christians from eastern and western Germany, being a split-up country with different currencies and government but united by the will to document to the world that the church of Christ was still a living power. And this demonstration did leave its impression upon all those who had gone far away from church and faith. During five days the sight of the city was stamped by the church meeting. In extra-trains and busses (2) and even with trucks (3) from all sides people streamed into the city. Big posters were greeting them (4) and all along the streets the flags with the violet cross were lined up. For all of them free quarters were given. Some Leipzigers slept on the carpet in order to have some guests with them in spite of the shortage of space they have anyway. In the mornings the streets were crowded with people pouring from the houses and aiming towards the place of the technical fair. Though one extra streetcar (5) was following the other, many people walked (6) because the cars were overcrowded, At the fair (18), they assembled in the many big halls where during fair time the big machines etc. were put up for the industrial visitors of all nations. All the big halls were thronged (7, 8, 9, 10), so people even had to stand outside to listen to one of the numerous loudspeakers scattered all over the area. (7, 11, 12). Also two big tents had been put up (13). In the tents (14) and in the halls mighty crosses were erected (8, 15, 16). From morning to night different topics were treated in the various halls and tents. It started with a meditation over a word from the Bible. Then followed an address about a special problem and in the afternoon a discussion about this problem. In one "Arbeitsgruppe" (studying group), the problem was the return of Christ. What we wait of Him and what He waits of us. How both things belong together: waiting for Him and working in this world for this world. What our life should like if it really was shaped by this waiting and not look like there was no other world, etc. In another Arbeitsgruppe the problem of the technic was discussed. We live in a world of technic; radio, airplanes, atomic power. Yes, our life itself has become technical, we have become driven like machines. The worker on his place is pressing one single handle all the day and starts to forget the sense for the outfit he is making a part for. Further, Technic is supposed to make the life easier, give people more time for themselves, but where are time, rest, meditation?

In another group they talked about the social problem. Have we Christians the duty to work for the better of the world or should we restrict ourselves to the mere religious ceremonies? What should be the relation between poor and rich, between owner of a factory and those who work for him? Why sends Paul the slave Onesimus back to his owner Philemon, but why does he write to latter he should keep his slave as a brother not only in religious concern but also in concern of the material needs (Psalm 15.16)?

Other questions arose in a Arbeitsgruppe where about the problem of peace was talked. Have Christians only to care about the religious peace with God? Or have they just kept peace in their families and with their fellowmen but not concerning wars? Should they leave the worldly peace to the politicians? Is the peace only a gift of God, so everybody who toils for it is betraying the heavenly grace? Is the war sent by God so we should do nothing against it; though we also call a doctor when struck by a sickness or prevent getting sick in spite of connecting also sickness with the almighty God? Well, I could go on enumerating all the questions having been ventilated. And it would mean to write a book to give a full account even of the one Arbeitsgruppe I took part in.

But one thing I have not to forget to mention, because it was the typical of the church-meeting at Leipzig. All the former meetings were either in Berlin or in western Germany (Stuttgart, USA zone; Hamburg, British Zone), this church-meeting happened to be in the DDR, where the official opinion of state and government is much in favour of the atheistic materialism. There were many difficulties and tensions between state and church caused therefrom formerly. And now the typical of this church-meeting was a new stepping-stone in the policy started last year, the will to go on together peacefully. The meeting was sustained by the state-officials in such way no church-meetings in western Germany were. This said the thousands of visitors from western Germany. The whole city was rendered. Even the one big hall built and owned by the Russians was offered (10). So it happened that the big cross was put up in front of the window showing a picture of the Kremlin, and beside the soviet insignia (15, 16). In the vestibule of the hall at the postal of the monument of Stalin, Christian nurses were passing (17), and behind the back the Christians were assembled in the hall (19, 8). Outside the memorial of the five-year-plan was surrounded by flags with the violet cross (20).

Besides also the theaters of the city, having just their vacations, were given to the use of the church, so at night Christian plays were performed by laity of Christian youth (21). And not only in the many churches of Leipzig but also in the big public movie-theaters Christian movies of western Germany were shown, what never happened hitherto. So the political impression of the church-meeting meant peace. And you will feel how glad we are about it. There is also hope it will help to ease the tension between East and West.

In order to give you a living impression of the big church-meeting I took some pictures for you. On their backsides I numbered them so you may relate them to the given report. You may show them around; there is even no reason against publishing a them—if not ill-used. On the lest Sunday of the meeting in the morning all services held in the big fair-halls, tents and in the hundred churches of Leipzig were that crowded a lot of people had to return home (22) or to listen to the loudspeaker transmitting the service of Bishop Dibelius (11, 12) who some time ago also was a guest in your states.

In our family life nothing has changed. Dorothee has grown quite a bit and is worshipped as queen by the other children of the village. Ehrenfried had quite a headache on account of the high-flood covering some of his land. My father is going on doing his schedule of work with all his love. And mother is worrying about the wellbeing of all of us. Everyone is healthy and hopeful. I am glad you gave such good note to the pictures I sent because photographing have become a hobby of mine and I am eager to improve. Perhaps there will come the time a Theologist has to have another profession besides to earn his living. St. Paul had to earn his living by making tent-clothes. I perhaps would take snaps. But for making it profitable until now a portable Electro-flashlight is lacking. There is none available here. And one needs it for good shuts. I wished I knew somebody by chance who has a used one but doesn't need it any more. Some of our reporters get it from western Germany, but since I am no professional with special tasks I don't get the different currency.

Well, in my next letter I will give you a new sample of my hobby. My wishes and prayers are with you,

Sincerely yours

Chris

(Postcard sent from Ostseebad)

August 18, 1954

Dear Miss Grover,

Hearty greetings from the Baltic Sea! A fortnight without books and writing, the time all filled up with swimming, sailing, sunbathing, and listening to concerts and plays at night. If I had a ship of my own I could bring this card personally, for the salty water goes from here to Florida.

At Leipzig and Dresden we received your gift-parcels! Many, many thanks! All arrived safely.

Sincerely yours and thankfully

Chris!

Four months has lapsed since the last letter from America.

November 20, 1954

Dear Miss Grover,

4 months passed since my last letter, a report about the big church day at Leipzig, and since 3 months I anxiously look out for the mail to find a letter of yours. I am more and

more getting restless. And I hope and pray there will be no serious reason. Each time I come home my parents ask about you and Mrs. Blakely and always I only can answer: No letter yet. In August also my father wrote you a few lines to thank you personally for your love. It was the time of summer and my vacation at the Baltic Sea, from where I wrote you a card. It was swimming-time, and now the time of skiing has come without hearing from you. We really are in sorrow about the American member of our family. Sometimes I think you might not have got my report about the Church-Day because I included some odd and unique pictures. Please, do drop me a note very very soon, only a short one.

Since summer I also sent you three packages and hope they have arrived undamaged and met you in perfect health. My oral examinations will be after Christmas because it takes such time for the professors to read my opus. At the time being I deliver a lecture about the epistle to the Hebrews and besides about the Synoptics (Matthew, Mark, Lucas) and about the so called Pauline mystic and its interpretation.

Please, do write soon! I will always remember you!

Sincerely yours

Chris

There are new restrictions on what can be shipped to East Germany. No canned goods are allowed, and only one package per person can be received per month. Christoph's dissertation is being reviewed. He notes the tenth anniversary of the firebombing of Dresden. While Christoph writes the death toll was "hundred thousands," the official estimates are closer to 25,000, which is horrible enough.

February 13, 1955

Dear Miss Grover,

Dear Mrs. Blakely,

I just listened to a sermon of bishop More of Atlantic City cast over the American network in Europe. It was about the meaning of the name of Jesus. Very distinctly spoken I understood each word of that excellent sermon, so it happened not to be as it sometimes does when listening to a show and I hear people applauding but don't grasp why. There were wonderful sentences in that sermon. As I ponder to remember one I just can exactly repeat one, he pronounced when speaking about the name of everything—though not making an evil thing good or a good thing bad since a name remains a mere name—but yet carrying a deeper sense of what the thing being named shall mean. The name of a baby he said embodies some ideas cherished in the heart of his parents. I was touched by the fine way he knew to set the words for the thoughts he wanted to express.

Now I guess I am to be blamed for letting you wait that long for an answer of your letter I welcomed that much when I got it before Christmas. Also, all three of your parcels arrived and spent a lot of holiday-joy. Each cup of coffee we drank meant a toast to you. And this I don't mean just allegorical, for since my family doesn't like to drink alcohol, though we are no abolishers (or as you call those people who want to abolish alcohol for complete abstinence) my parents only take a drink when they think it impolite to decline. The younger generation of us (we three brothers) are not of exactly the same strict way. So our drinks at home consist mostly of tea and—after the arrival of a parcel from you—of coffee. And though we don't touch each other cups (in the way toasts are done) when sitting around the table and remembering you and wanting you happiness and health. I really can say that during the last night of the [last] year you did get some toasts with the additional speeches. And when midnight came and the start of the new year, I had to remind my parents that you still lived in the old one because you would step into the new one 1955 six or even eight hours (as far as I remember you have two different times in the states) later. So it might have been just the hour we breakfasted before the morning service at New Year's Day when you shouted "Happy New Year!" also over the ocean. I think our hearts will have met and shaken hands. And my wishes and prayers are with you, at the last day of the by-gone year, and at the first day of the new one and at every one.

A reason for waiting with this letter were the restrictions for import and export pending to be issued. Now there is cleanness about them and I am glad they don't concern the little gifts I had bought at Christmas-time for you. [Having been delayed], they will go now this week. But for the parcels you sent there is a new restriction I have to inform you about. No canned food is allowed anymore because our government is afraid of weapons or imperialistic propaganda material shipped in hidden within the cans. So I read in the newspaper. All the other things are not changed by the new regulations. All things enclosed by glass, cartoons or paper are not concerned, also if cans are [opened], they are allowed. So for example: Lard in cans is forbidden, but if it is packed in paper or so, it is allowed, also if the container is opened. People here are not at all happy about this regulation. And there is another restriction besides. Each person is allowed to receive only one package or parcel from western Germany or from abroad. If there comes a second within one month it will not be handed over. That won't concern me because I [only] get parcels from you. Only my father as a clergyman sometimes gets packages from western Germany to distribute. So there would be the slight possibility that your parcel would be the second within a month and so not handed over. Because he is a clergyman his address is in the clergymen-book and people of western Germany who have no relatives in the DDR but want to send some gifts for needy persons sometimes pick addresses from this book and send worn garments and some food to pass to a needy person. Sometimes at Christmas he all of a sudden gets a package even for himself from a person completely unknown to him. So bridges go from one part of Germany to the other one. And they are broken now. Because people in western Germany know about that restriction they will stop sending, everyone afraid to be the second to send to a person and so in many cases nobody will do. But there is no reason to despair. Since each member of a family may receive one parcel each month my father asked me to pray you to take instead of his name the name of my mother or even better the name of an elderly aunt of him who is living with my parents since some time because she cannot stay by herself any more. Her name is "Fraulein Ida Haufe." Please don't be troubled by those new regulations; perhaps they will be lifted someday soon. The tension between the countries always seem to carry along a tension stimulated for the people's relations between them. So I hope the governments of this world at last will come together and settle their problems peacefully so a time will come which doesn't try to split us but leads us even more together than possible by mail.

Today is the day ten years ago Dresden was [destroyed] within one night. No house was left [standing] within the city. All was a fire of flames, even the streets were burning by phosphor and those who could escape from the cellars of hit houses turned to living torches running till they stumbled dead. Hundred of thousands died in one night. The city herself died. The leftovers were a single heap of debris and some chimneys pointing towards the sky, as the towers of the churches did, the meaning of which people forgot by waging that war. I have to remember this night of Dresden's death today very intensely, as you will understand, and I am stunned by all this new war-noise in the world. Why cannot be all people friends as we are, and help each other. It would be much better for everyone and the kingdom of God would be nearer besides the coming of which we pray for. The Lord may bless our friendship and all the love which combines the countries and strengthen them in such way all those who sow hate will be overcome by love.

My oral examinations are still standing out. Since all professors have to read my three volumes first there might come spring, I fear. But our Decan promised me he will do all he can to end the promotion this term.

Our family saved money through the whole last year in order to buy a [tape recorder]. At Christmas we had it and it gave us a lot of fun. The first time we could listen to our own voices and nobody could recognize his own, though it was recognized by every other one. Also good music broadcasted we can put on the tape and so replay it at any time we want. Now we had a great idea. If you have the opportunity to play such tape we all would speak upon one and send it to you. If it is allowed to send, I still have to inquire. The tape runs with a speed of 9 centimeters (that is a little more than three and a half inches I guess) in the second. So you could hear our voices.

At Christmas I also took some pictures at home. These I will send you in an ordinary letter. All of my family asked me to greet you and forward all their good wishes. I also want to thank you for all you said and did.

Sincerely yours always
Chris

June 24, 1955

Dear Miss Grover,

Dear Miss Blakely,

Before I start to leave for my vacation, I am happy to thank you for your letter and both the parcels to Leipzig and Dresden. The parcel for my parents just arrived when they had left to spend their furlough in a very small village of western Germany near the Denish border. It is the first time after the war they went to the western parts of Germany. You remember it was nearly impossible for the first time to cross the zone-border splitting up our country, but [for] two years those difficulties had been lifted. Nevertheless for those who have no relatives in western Germany it was impossible to visit those parts of our country because we still have two different currencies and people are not allowed to take money with themselves in order to exchange it. This year the church of western Germany invited some reverends of the DDR to spend their vacations in the homes of their western colleagues, so my parents went via Hamburg to the small village Ulsnis. They wrote it is very lonely there but just the right for recreation. And that both of them need badly. During the past months they did ask in each letter from home about you and whether you had written. And when at Easter and Whitesunday I was staying at home we often talked about you. At Whitesunday we drank the last cup of your coffee. So you may imagine the joy that arose when receiving your letter and shortly afterwards your parcels. We really don't know how to thank you for not forgetting us. The parcels arrived in perfect condition and they completely corresponded the new regulations. Only *air-tight* closed *tincans* are not allowed, so tea and cocoa may be left wrapped up as they are. I am so sorry that you have to undergo this trouble of paying attention to those changed difficult regulations, but I am so happy they didn't keep you from remembering us.

This year I took vacations for a fortnight and will go to the Baltic Sea together with my brother John. We will be on Ruegen island and hope the weather being rather cool until now will change for the better so we will get the opportunity to do a lot of swimming. John is still a clerk in the church-office administrating a big grave-yard. He hopes to pass the examination for inspector this year or next. When being on vacation and asked for his occupation, he always hesitates a little bit because people don't like to be reminded of a grave-yard in the middle of their vacation-joy. But to some it certainly does good to remember life is going to an aim and not always running on.

Only Ehrenfried and his wife cannot take any free days. For their garden there is now the high season and the work doesn't end. They never sleep longer than 4 hours a night. Hands are very rare and so most of the work has to be done by them themselves. But they are going on; the last winter they could build a fine green-house and so are able to grow early vegetables for which they get a good price. Dorothee is rather a darling. At Easter I went with her to a fair and she had a lot of fun. She had to stay with my parents for quite some weeks because Hertha had been in the hospital. My parents liked to have her with them but when she parted they said: it is right that the young people have the children. I include some pictures I took to give you an idea how much she grew up since last year. My last letter with the pictures was returned because its weight was too high.

In the meantime, I passed my oral examinations and concluded my promotion on April 28. I am very happy this heap of work has been done. Everything went smoothly through two or three questions I couldn't answer to. But I was examined for over two hours. At first we talked about the problem "realized or futuristic eschatology," then I was asked for the judgment of the New Testament and the early church on slavery compared with the opinions of the Jewish, Greek, and Latin world. The last question concerned the latest developments of "Psychoanalysis." I got the degree *sum laude*.

I am glad you liked the little gifts of mine. It gives me a thrilling feeling to imagine that things I bought an held with my hands and wrapped them up are now standing in front of you. The novel *Bambi* I don't know. But I remember to have read a novel called *The Yearling* and it must be somehow similar. I [haven't] read much English during the last two years and I feel the result. But now I hope to find the leisure time for this occupation again. I will stay at Leipzig. Next term I will render a lesson about the specialties of the gospel by Lukas.

I hope to hear soon from you. My thoughts are always with you. My best wishes and prayers for your health and happiness.

Sincerely yours as ever,

Christoph

Christoph merits three mentions in the press—December 17, December 20, and December 30. The first two cover the same topic, an article Christoph has published in the Protestant monthly magazine *Faith and Conscience*. The last one discusses the Disciple Paul. Christoph is described as postdoctoral candidate for New Testament theology.

There is a one-year gap here. My guess is some letters simply got lost. Christoph is excited to be going to Frankfurt for a church conference. He is settling into his career and doing well as is evidenced by his invitation to the Theological Seminary in Prague.

August 2, 1956

Dear Miss Grover,

The two parcels of you to my parents and me safely arrived. We all long to thank you for this precious and welcomed gift. My father wanted to write you his thanks specially, because he most of us all needs the coffee against his headache and so suffers when there is none.

Tonight my train will pull out of the station of Leipzig to carry me to Frankfurt at the Main-river, Western Germany, where this year the church-day of Germany will be. Thousands of Christians will be assembled there. I guess I wrote you about the big church-day we had at Leipzig three years before. The whole city was crowded by people carrying the badge of the church- day and going singing through the streets, an overwhelming impression of the mighty membership still clinging to their saviour. From the DDR special trains go to Western Germany and many will go to come together with their Western German brothers and sisters. I am glad to take part at this meeting and visit the addresses and discussing what will be about the different problems of interest. I wished we could meet there at this occasion.

In June I was invited by the Theological Seminary of Praj. So l went with 13 students of theology to our *[missing]*. We were welcomed very heartily; and through a fortnight we lived together with the students in their home. Each of us stayed together with a theological student. We took part in the lessons, which were rendered by their professors. And by truck, bus or train we went to distant cities and towns where protestant Christians lived and visited their assemblies. We also saw historical towns like Tabor, where the Hussite movement started. The houses of this town all are combined by underground tunnels, built for defense purposes. It was like a labyrinth. We also saw the new village of Lidice. The old one was completely destroyed by the Nazis during the war; it was a revenge for the attempt on Heydrich, the German protector of the Checks. All the men were killed, and the priest, who was offered to be left alive, asked for the same fate as his brothers suffered. The women were brought to concentration camps and the children taken away. They are not all found out where they are. The debris has been carried away and the place was supposed by the Nazis not to remind of this village. Today only small posters show where the church and the school were. When we went to Czechoslovakia we wondered how we were welcomed, because we know how much the people had suffered before and during the war. And we were surprised by the heartiness we were welcomed with. Nobody on the streets was angry to hear German voices, but many asked us where we came from, etc. When we asked for direction (we often went alone through the city) and then asked one didn't understand German; another passersby stopped and asked us whether they should interpret. We all were glad that brotherhood and friendship between our people was possible again.

Naturally we had to see a lot at Prague, the wonderful Hradschin castle, and the many beautiful churches. Most of the people are Catholic, because during the counter-reformation the Protestants were Catholicized. After the Bill of Toleration (1280) *[I think 1280 would be a bit early since it predates the Reformation; I suspect Christoph is referring to the Edict of Torda (1568)]* not many remembered the protestant faith of their forefathers. The sign of the Protestants (and also of the Czechoslovakia Church, which separated from the Catholic one after the first world war) is the cup (the cross is more the sign of the catholic ones), it remembers of the claim, not only to give the bread of the Lord's Supper to people but also the wine.

The time we were at Prague, just the musical festival had started. So we had the opportunity to listen to a lot of good music. Three times I visited the opera. I saw two operas by Czech composers (Janacek and Swetlana) and saw a wonderful ballet (Swan Lake) by Tchaikovsky. When I was asked at the end of our staying for something I was not content with, I could only say, that this was the short time of sleep we had, but this was necessary, because so many things were shown to us. At Tabn I met an evangelical minister who had studied in the USA and married there. At the time being his wife is with their parents in USA on visit. We could have a discussion in fluent English. Many speak German. All theology students have to learn it, in order to be able to read the theological German books. There were many problems, theological and other ones, we had talked over.

Some snaps I enclose. One is showing me crossing the Woldam [possibly the Vltava, which Germans call the Moldau and which is commemorated by Bedrich Smetana in his composition *Die Moldau*] where it is a big reservoir. On two you see me at the memorial of John Hus. On the other one I am with the students standing on the famous Charles-bridge, behind us the Hradschin and St. Vert. On another snap I am studying the Hebrew inscriptions on an old Jewish cemetery where I found a stone-sarcophagus. The other pictures are about home. My father, my mother and me. The one shows my mother just making one of her wonderful cakes. Do you recognize your gifts on the table? With many hearty greetings,

Sincerely yours,

Chris

POST CARD

A postcard from Frankfurt.

September 6, 1956

Dear Miss Grover,

The best greetings from the big meeting of the German Evangelical Church in Frankfurt, Western Germany. Thousands of Christians from our whole country met here and discussed. about the problems of our present living. All of us had the opinion that the one should come now our divided country should be reunited. I often remember you, especially each time I see an American soldier here (Frankfurt belongs to the Zone occupied by the Americans). At the church meeting there also are some visitors from America.

How wonderful if we could have met here.

Sincerely yours,

Chris

POST CARD

Christoph's status grows. He is invited to Sweden by the Quakers, probably as a result of his association with Emil Fuchs. This postcard and the following one are dated only "1956."

1956

Dear Miss Grover,

The best greetings from Sweden! I was invited by the Quakers to take part in an international meeting. I am sorry there is no American here. The meeting takes place at a wonderful lake surrounded by mountains and forest, so we take long walks. There are only some small wooden houses we live in. Next week we will go to Stockholm and stay there for some days. We are impressed by the high standard of living people have here.

Sincerely yours,

Chris

POST CARD

1956

Dear Miss Grover,

From the mountains of Thuringia with their forests covered with snow I want to give you my kindest regards. It does a lot of good to have all day fresh air and to get strengthened this way for the tasks of the new term. With 25 students there are a lot of discussions too.

All good wishes for you and your family, from

Chris

Tabarz/ Thur, Haus Fortuna, DDR

Christoph's career is taking off. He's mentioned in the September 7, 1957 edition as having led an ecumenical holiday camp that featured young Germans, Czech, and Hungarian theologians in conversation and again in the January 18, 1958 edition discussing the publication of his book, *Die sittliche Rechtfertigungslehre des Paulus* (Pauls' Moral Doctrine of Justification) (Halle Niemeyer, 1957).

TRANSLATION

This letter, from Christoph's mother, is a translation—probably by Christoph. The gaps between letters are growing longer.

Dresden, July 3, 1958

Dear Miss Grover,

I promised Christoph that I would write to you. Now I must keep my promise. Unfortunately, my school knowledge no longer suffices to do so in English. However, perhaps it will be fun for you to get a little practice in translating. And you will surely succeed, as you did once before.

I should like, on behalf of the entire family, to thank you very sincerely for your valuable packages. They always provide real pleasure and benefit. May God bless your great love, your kindness, the trouble you take, and not last, your sacrifice. I myself sent you, about ten days ago, a little token of thanks. In addition to the organ you will receive next time a couple of angels just like the ones who are playing. Do you still continue to enjoy these little things?

The book markers I bought from the "Homeland Messengers." This is a group of people who warm the hearts of men with genuine folk art in word and song. Their leader is a quite charming and gifted woman and is carrying on the work of the late Chancellor Seiffert; he was the founder of the Provincial Museum. His widow is still alive and is 94 years old. She recently attended the twelfth anniversary celebration of the group in fairly vigorous health. It was another lovely evening. The motto of the "Homeland Messengers" is: We want to make people happy. And that is what happens every time they appear.

Now summer has come, and vacation time is here. It does everyone good once a year for a short time to take a rest. We shall this time probably go to the seashore. Next year, God willing, we want to take our 7-year-old granddaughter with us. Her parents, as garden workers, are busiest in the summertime and are never able to go away, and; a change of air will certainly do the child good. She has to study hard in school and in her free time take care of her 1 1/2-year-old little brother. We take great pleasure in the two children. They are both chubby-cheeked and very active mentally.

I myself now have a lot to do in our rather large garden.

Since I can no longer have any help, the house, garden and the many things that my husband's position entails for me, too, are often a bit too much. But God has so far given me the strength for all my work.

We are always glad, through Christoph, to hear from you and are glad to share in your nice family life.

We wish you a nice vacation time. May God bless you and all your large family.

With thanks, I remain

Sincerely yours,

Edith Haufe and her husband

P. S. I am taking the liberty of adding a special cordial greeting to Mrs. Quincy Blakely.

Yours devotedly,

Ehrenfried Haufe and his wife

The June 29 edition mentions the professorship for New Testament studies was filled by Christoph.

POST CARD

A postcard from the Baltic Sea.

July 20, 1958

Dear Miss Grover,

I want to send you some greetings from the Baltic Sea, where I stay for a fortnight vacation. Weather was quite good for this region, so I am rather dark-brown, though it didn't go without a little peeling. During the rainy days and in the evenings I do some preparation work for my lectures of the new term. With best wishes to all of your family,

Sincerely yours,

Chris

Binz/Rugen Haus Cranitz DDR

Christoph continues to make the newspaper. The October 18, 1958 edition again notes Christoph's filling the professorship for New Testament studies. The October 31 edition as well as *Neues Deutschland* and *Berliner Zeitung* on the same date note his (as well as Emil Fuchs') signing the Magdeburg Declaration, which calls for, among other things, resistance to nuclear armament, and endorsing candidates of the National Front.

POST CARD

Again, another gap between communications.
Christoph writes from Albania.

1959

Dear Miss Grover!

From an interesting country I send you my best wishes and greetings. Here (Albania) the water is warmer than at the Baltic Sea. I am getting brown very fast. But most I like to meet people here, which nearly all are Mohammedanians, because Albania was for 500 years under the Turks (until 1912). The country lies between Jugoslavia and Greece, but is very different from both, also concerning the people. There are wonderful motives for my camera. Well, I'll give you a report when having returned home.

Sincerely yours,

Chris

Christoph is mentioned in the March 8 edition as taking part in a pulpit declaration regarding the stationing of nuclear weapons in Dortmund, West Germany.

Christoph appreciates the instant coffee he is receiving from America and describes his workload. His younger brother is getting married, and his family would like to see Christoph do the same. Note Christoph, who is now thirty-four, has a realistic view of children and seems quite content to remain an uncle.

May 2, 1959

Dear Miss Grover,

That was a big joy to receive a parcel from you again! And the Maxwell Coffee "New, Fresh-Roasted Flavor" is delicious! I had run out of instant coffee since quite some while, so it was a special enjoyment. This term my lessons start early in the mornings, so I am always quite in a hurry to catch my tram. But the instant coffee is really made instantly! And so I arrive at university quite awake. The early lessons have some profit though. At noon I am home again already and have the rest of the day for studying. Beside the lessons there are quite a lot of obligations. This term I had to examine a rather high number of students who have finished their studies. Each student has to render a treatise of about 60 pages about a special theme, for which he had a time of 6 weeks. Besides in seclusion he has to write a three-hours work and afterwards undergo an oral examination. Furthermore, I had some dissertations to correct and vote. Another one I will read over Whitsunday when staying home. Strenuous all the reading is in case no new ideas can be found but only commonplaces. Then it's hard to concentrate and not to think astray in dreaming. The term ends 20th of June. Two days afterwards my brother John will get married. So of our family I am the only one staying single yet. Well, it's good the children have an uncle to whom they can come and who isn't worn out by children's bawling yet. Here in the house all the families are so sorry I am not married, because—as they say—I am "so fond of children." (Their children sometimes come and ask for a candy). But I think perhaps fondness would be worn out when constantly surrounded by them. Mother says: having little Christian around for one day is more strenuous than two hard washing days, because he always claims for some new entertainment, otherwise he will destruct anything. After the wedding of my brother, I will go for a fortnight to Albania at the coast of the Adriatic. It will be the first time in my life that I go somewhere by plane.

I hope, the new package I send, will go safely to you. All best wishes to you and your family.

Yours thankfully,

Chris

Christoph is again in the news. In the July 18 edition he's noted as having traveled to Vienna for the Vienna Youth Festival.

This is the final letter. There may have been more that have been lost. Food was no longer rationed, and it's possible Christoph became so involved in his work that he simply stopped writing or that his American benefactors noted he could afford to fly to Albania and probably no longer needed their help. Another possibility is Klaus Fuchs was released from prison in England on June 23, 1959, a few weeks before this letter, and settled in Dresden in September. Perhaps Christoph felt conflicted about corresponding with

Americans after Emil's son's ordeal. Perhaps it's as simple as correspondence with Americans was discouraged.

August 15,1959

Dear Miss Grover,

Thanks for your letter I had taken home!

Now the weeks of travelling are over and I have to start to prepare myself for the new term, during which I have to render a new lecture about Paul's letter to the Romans, four hours a week. Many interesting things happened during the last two months. John's wedding was wonderful. I think father has written you about the festivities and sent you some pictures. During the—ceremony—celebrated by my father—when he asked the questions tears were in his eyes. After a vacation in the mountains (honeymoon) they live now together with my parents. They had bought new modern furniture for a living room, the sleeping room will be furnished later on when they will have saved the needed money.

Very interesting the weeks in Albania turned out to be. It's the poorest country of Europe. For 500 years they were under the Turks till 1912, so they always were exploitered and practically belonged to the Orient, so there were nearly no connections with Europe and therefore also no industrial development. Most of the population (70%) are Mohamedanians, some Greek-orthodox (20%), and only a few Roman Catholics, mostly in the North, where the Venetians (Italians) had [conquered] some parts during early pasts. Mostly it was difficult to take snaps of the population, because as Mohamedanians they declined of making pictures of men. Children were more ready, but often blamed harshly by older people seeing their readiness. Nevertheless, times change. Until some years ago the women went veiled; besides they had to do all the work, while the men loitered around.

He rode on the mule and she had to walk behind carrying the luggage on her head. Now the men are educated to write and work, streets are built through the mountains (I never drove such dangerous slopes) and the rich mineral sources of the country are getting [mined]. For the tourists they do everything, drive us around the country, giving us the best food and rooms of the few bearable hotels, because for this they can buy machines, etc. I enclose some pictures to render you an impression.

Sincerely yours,

Chris

After the Letters

This chapter was a challenge. So many people who had read the letters before I began work on the book felt like Christoph had become a friend. Many asked what became of him, had he married, etc. I simply had to say I believe he became a full professor at the University of Leipzig, but I'd been unable to verify even that. My letters to the university were ignored at first, and the last one curtly replied, "Unfortunately Prof. Huber doesn't have any information about Haufe."

While the draft was being reviewed, I decided to try one last time to track down the elusive Christoph. Nancy Thorndike Greenspan had written in her biography of Klaus Fuchs that Emil Fuchs' grandson, Klaus Fuchs-Kittowski, had traveled with him to America and England in the late 1940s and had become a student at Leipzig in 1956. I thought there might be a chance that Fuchs-Kittowski had known Christoph. I was able to locate his son, Frank Fuchs-Kittowski, who at first thought I was asking about Christoph Michael Haufe (1930–2011). Frank also informed me Emil Fuchs had left Leipzig for Berlin in 1959. When I explained the Christoph I was researching was not Christoph Michael and included some excerpts from Christoph's letters discussing his involvement with Emil Fuchs, he contacted his father, who asked him to send me his email, saying, "My father knows a Christoph Haufe from Leipzig very well. But I'm not sure if you mean the same one." At first, I interpreted this to mean Christoph was still alive and perhaps I could get him to clear up some questions I had about his letters, and I responded with an enthusiastic email. So far, I have not received a response. Klaus Fuchs-Kittowski turns 90 this year.

At the same time, I found a pamphlet online for a 2002 Walter Markov Colloquium on Christianity, Marxism, and the Work of Emil Fuchs. Klaus Fuchs-Kittowski was one of the presenters, but most helpful was a footnote that credited Christoph with having edited the first (1952) edition of Emil Fuchs' *Marxismus und Christentum* (Marxism and Christianity). It gave Christoph's birth and death dates as 1925 and 1992. This information took me to a page for the University of Leipzig's Institute of New Testament Studies, which tells the story of the Institute. It begins:

The theological faculty is one of the founding faculties of the University of Leipzig. The interpretation of the New Testament has been taught at our university since 1409. In its long history, the Institute for New Testament Studies has produced some very important researchers, including Constantin vin Tischendorf and Caspar René Gregory. During the dictatorships, there were also controversial figures among the professors. We introduce some of them below.

One of the controversial ones was Christoph:

In 1955, the death of Oepke and the retirement of Leipoldt [Christoph's predecessors] created a difficult situation. There were attempts by the state to influence the successor. Efforts to appoint Ernst Bammel (later in Cambridge) failed. Leipoldt represented himself until 1959. Meanwhile, the State Secretariat for Higher Education and Technical Education installed Christoph Haufe (1925–1992). Haufe was an employee of the Ministry for State Security ("Stasi") as a secret agent (GM), code name "Blume." However, against the will of the majority of the faculty, he was appointed lecturer in 1957 and professor in 1958. He remained in this position for 26 years as the only full professor (chair) for New Testament.

So Christoph remained in his position until 1984, when he would have been 59 years old. He died three years after the fall of the Berlin Wall.

All of this explains somewhat why so many people, including university personnel, denied having any knowledge of Christoph. He was an embarrassment. It also explains why the letters stopped. Between his pay from the university and the Stasi, Christoph no longer needed support from America.

I finally hit pay dirt with Dr. Jens Herzer, who is the current head of the Institute for New Testament Studies at the University of Leipzig. He referred me to two people who had known or known of Christoph, one of which had been a student of his.

Dr. Jürgen Ziemer wrote: (This is a translation via Google Translate)

"Excuse me for the delay. I had to wait until I heard from my colleague Prof. Hartmut Mai, whom Christoph Haufe had studied with personally and who knew him until he died in 1992. Christoph Haufe studied theology in Leipzig, including an exam as a translator in English. In 1955 he studied at the Theological Faculty of the former Karl Marx University as Dr. of Theology, Doctorate in New Testament; 1957 habilitation in the same subject. From 1968 Haufe was professor of New Testament and Jewish studies. In the year 1984 he retired for health reasons. Theologically Haufe was primarily concerned with Paul (doctrine of justification). Politically, there was a certain proximity to the socialist state of the GDR. Haufe was very connected to Prof. Emil Fuchs.

"Christoph Haufe was married for the second time. I don't know if they had children."

Dr. Wolfgang Ratzmann wrote: (Again, courtesy of Google Translate)

Dear Mr. Roth,

"I experienced Christoph Haufe as a professor of New Testament theology from 1965–70. He was not popular with us students at the time. His explanations of the texts were boring and without any reference to us today, his lecture soporific.

"He had come to his professorship through the SED, not through his theological competence. Christoph Haufe was an unofficial employee of the GDR "State Security" (Stasi). We knew that he was close to the SED. And in this he differed from his younger namesake Günter Haufe, who was then an assistant for New Testament in Leipzig, who introduced us well to the historical-critical method of exegesis and preached convincingly himself. Günter Haufe later became a professor of New Testament in Greifswald.

"Christoph Haufe probably married late. But he had no children. In retirement, he completely withdrew from the faculty and from the church-theological public. I don't know anyone who had any contact with him."

Both emphasized Christoph is not to be confused with Christoph Michael Haufe, who became honorary professor after a lectureship at the Theological Seminary in Leipzig.

Phil Leask was reviewing my draft when I found that Christoph had worked with the Stasi. He had already found indicators that Christoph was working with the East German government. He wrote that it was a treat "to be able to read the letters over time and see both Christoph Haufe's personal development, the changes from his early days in the Soviet Zone of Occupation, then in the GDR, and the way the GDR itself changes. He is clearly a very intelligent person with ideas of his own about religion and about the link between religion and society and that was also wonderful to see these American women packing up and sending all these things he needs, initially to survive and later to have as luxuries.

"Then we see his career taking off and him attending things in foreign countries.

"That always rings alarm bells. The Stasi had many ways of getting people to work with them—rather than directly for them. People in prison, for instance, could find themselves offered early release and a more comfortable life if they agreed to act as an IM - Inoffizieller Mitarbeiter, an informer not directly employed in a full-time role but reporting regularly to a minder or handler. By the end there were huge numbers of these, so everyone in every workplace or organization or group or, of course, church, would have someone around them in this role.

"Another way to get someone to work with the Stasi was to see where their personal views or activities could be used to help the aims of the GDR. The bait was often that the person would be allowed to travel to the West to take part in events or even to address meetings or conferences on things that were important to them.

"Christoph was passionate about international friendship and the need for peace. His strongly expressed views seem genuine. The GDR was also keen on peace, as long as it was in line with the wishes and needs of the Soviet Union. Hence the desire—which Christoph endorses—for a neutral and united Germany.

"So a possible scenario is this. The Stasi, knowing he is constantly writing to people in the US and getting parcels from them, and having read his letters, see his desire for peace and his thoughts on neutrality. They see he is a hard-working pastor, mixing with 'ordinary' people and listening to their complaints. They also see he is intelligent and could benefit personally from doing more study. Theology is not to their liking, but there he could be useful to them by reporting to them on those around him, by giving them a clearer understanding of what the churches want and how the Stasi can turn this to their advantage, and by supporting causes such as the peace movement—close to their heart. So they make sure Leipzig University offers him a place and funding to do his doctorate. In return, he agrees to report to them. In return, as his standing is raised and he becomes more prominent, they are happy for him to become a professor. On it goes. Soon he is going to West Germany, to Sweden, to Albania ... he talks about peace and international co-operation. He criticizes the West for the way it treats his counterparts there who support the GDR's efforts to get peace on Soviet terms. He enjoys his life. He has privileges. He becomes a recognized public figure.

"What makes me think this? Firstly, it was a common enough trajectory. Secondly, I have done a quick search of the three main national papers in the GDR, *Neues Deutschland, Berliner Zeitung,* and *Neue Zeit.* There are around seventy mentions of him in reports, the earliest I could find being 1955, a relatively 'harmless' one about him giving a speech about one of his professors. In 1957 he is leading a camp for like-minded Christian theologians from the GDR, Czechoslovakia and Hungary, and they are enthusiastic about what the socialist regimes are doing in the East. On October 31, 1958, he is a joint signatory with Emil Fuchs and others in support of the SED's efforts to build socialism, with specific points about peace and stopping West Germany having atomic weapons. They commit themselves to voting for the National Front candidates; that is, the single list of candidates from several parties but all controlled by the SED, the communist party. In other words, by 1958 he is clearly combining his theological activities with support for what the GDR is trying to do. There is from then on a particular focus on peace, but that also sees him speaking up for the GDR regime and against things being done in West Germany. And then he travels ...

"So I might be wrong, but it looks as though he was recruited in the mid to late 1950s. It could have been earlier, leading to him doing his doctorate."

Phil also brought up the question of Christoph's involvement in World War II as well as that of his father's. Christoph would have been 20 when the war ended; Ehrenfried would have been 49. Both were well within the age range of those being drafted in the final days of the war. In his book, Phil writes of a battle that took place near Schönebeck in April 1945 in which fifteen- and sixteen-year-olds as well as old men were forced to fight with predictably tragic consequences.

Ehrenfried is listed as the minister at Dresden-Breisnitz Evangelical Lutheran Church from 1923 to 1947. As for Christoph, that will have to be an unanswered question. Wolfgang Ratzmann said Christoph never discussed his private life.

I really don't know how to bring Christoph's story to a satisfactory conclusion, so I'm going to have to settle for an unsatisfactory one.

Living in East Germany undoubtedly required many compromises. In his introduction to the third volume of Victor Klemperer's diaries, which covers the years 1945 to 1959 and is titled, appropriately, *The Lesser Evil,* Martin Chalmers writes, regarding the Klemperers' decision to join the Communist Party:

"This sounds as if a degree of opportunism was involved. But, then, what apparently purely political decision in Germany in the early postwar months and years might not seem in part and in retrospect opportunist? Every such decision was simultaneously about how and where best to obtain food, to keep a roof over one's head, to find some way of earning a living."

Phil Leask's phenomenal *Friendship without Borders: Women's Stories of Power, Politics and Everyday Life across East and West Germany* tells the stories of a group of 29 women who were classmates whose letters from the late 1930s to the late 1990s demonstrate the privations, challenges, and, yes, compromises everyday people dealt with during challenging times on both sides of the border. I didn't live those lives, nor did I live Christoph's life. I don't feel qualified to make a judgement on his decisions.

Further, please consider that every day millions of people are asked to compromise their values in order to "obtain food, to keep a roof over one's head, to find some way of earning a living." Sometimes these compromises have disastrous consequences, as we have seen in the case of the Boeing Company's profit over safety practices.

I would be happy to know my readers' thoughts on these issues, and I'd be more than thrilled if anyone who reads this book has any additional information on Christoph. Please feel free to email me at live_cheap@att.net and put "Haufe" in the subject line. With the caveat that I'm past my biblical sell-by date, if I'm able, I'll respond.

Christoph Haufe in the Media after 1959

1960

In the January 14 *Neue Zeit,* Christoph, along with Emil Fuchs and others who are editors and staff at the monthly magazine *Faith and Conscience,* speak out against the trivialization of antisemitic excesses.

On January 30 the same publication notes Christoph's inaugural lecture at the Theological Faculty in Leipzig. The topic was "The Ancient Assessment of Slavery."

1961 (The Berlin Wall went up August 13, 1961.)

On September 16 the *Neue Zeit* notes Christoph took part in the International Congress for Free Christianity in Davos, Switzerland.

On November 14 the same publication shows Christoph attended a meeting in Gorlitz, East Germany to discuss a German Peace Plan, which was attended by "prominent guests from abroad." The *Berliner Zeitung,* reporting on the same meeting reports Christoph stated pastors and laypeople are obliged by the Gospel to accept the socialist state in general and work with it to achieve what has been recognized as the common good.

1962

On January 24, East Germany introduced general conscription. All men between the ages of 18 and 26 were required to serve eighteen months in the National Peoples Army. There were no exceptions for conscientious objectors. The January 29 *Berliner Zeitung* quotes Christopher as follows: "The responsibility for peace also requires Christians to support all measures that can prevent the outbreak of a third world war."

The July 18 edition of *Neue Zeit* mentions Christoph is

in Helsinki proclaiming Christians have a political responsibility to actively participate in maintaining peace in the age of the atom bomb.

On September 6 the *Neue Zeit* notes Christoph has responded to numerous questions from West German guests of the Christian Democratic Union in Leipzig.

The September 29 *Berliner Zeitung* reports Christoph participated in a meeting in Leipzig on the topic of "The Peace Service of Christians."

1963

The June 6 *Neue Zeit* reports Christoph, along with Emil Fuchs and many others calling for peaceful coexistence.

The August 28 *Neue Zeit* reports Christoph and twenty-three others signed an open letter to West German clergy regarding the War Crimes Trials saying the punishment was not adequate.

In the October 17 *Neue Zeit* and the *Neues Deutschland* Christoph is reported to have joined many other theologians in calling for support of the National Front in the October 20 elections.

The December 7 *Neue Zeit* reports Christoph's appointment to the chair for Modern Theology.

1965

The September 25 *Neue Zeit* notes Christoph has been confirmed as Vice Dean.

The October 7 *Neue Zeit* reports Christoph and several other theologians, pastors, preachers, and church council members gathered for a discussion of "Christian responsibility and civic decision."

1967

The January 28 issue of *Neue Zeit* reports Christoph took part, along with other teaching staff, in a series of "Christian Circles in the Committees of the National Front" working groups discussing confronting social problems from a Christian perspective.

1974

The May 18 *Neue Zeit* reports Christoph took part in a commemoration of the twenty-fifth anniversary of the GDR.

There's no way to know why media coverage of Christoph ended a decade before his career did. Phil Leask wrote that the SED may have decided he was no longer much use to the Stasi and the SED, so they moved on to someone else. He also wrote the lack of mentions could mean they were no longer going to support him and send him to events. Or it could also be that his cover had been blown and no one at Leipzig or anywhere else trusted him anymore.

The last may be the most likely and is consistent with what Wolfgang Ratzmann, Christoph's student from 1965–70, wrote.

Bibliography

Bergen, Doris L. *Twisted Cross: The German Christian Movement in the Third Reich.* University of North Carolina Press. Chapel Hill. 1996.

Fulbrook, Mary. *The People's State: East German Society from Hitler to Honecker.* Yale. 2005.

Greenspan, Nancy Thorndike. *Atomic Spy: The Dark Lives of Klaus Fuchs.* Viking. New York. 2020.

Hoyer, Katja. *Beyond the Wall: A History of East Germany.* Basic Books. New York. 2023.

Klemperer, Victor. *I Will Bear Witness: 1933–1941.* Modern Library. New York. 1999.

———. *I Will Bear Witness: 1942–1945.* Modern Library. New York. 2001.

———. *The Lesser Evil: The Diaries of Victor Klemperer 1945–59.* Weldenfeld & Nicolson. London. 2003.

Leask, Phil. *Friendship without Borders: Women's Stories of Power, Politics, and Everyday Life across East and West Germany.* Berghahn. New York. 2023.

Acknowledgements

This book would not have happened without the help of my good friend and former University of Missouri-Kansas City (UMKC) professor of political science, Max J. Skidmore, who saw the value of Christoph Haufe's letters as primary source material, recommended the project to his publisher, Westphalia Press, and went well beyond the call of duty by reviewing my first draft while on vacation in London.

Hearty thanks go to UMKC Professor of History Andrew Bergerson for recommending sources for studying life in postwar East Germany and especially for referring me to Phil Leask, a writer and researcher at University College London. Phil agreed to review my draft, and when I wrote to him about Christoph's activities after 1959, he took the time to track him through German newspapers. This book would not be the book it is without his input.

Thanks also go to Jens Herzer, head of the Institute for New Testament Studies at the University of Leipzig for referring me to Jürgen Ziemer, who knew of Christoph and Wolfgang Ratzmann, who had been a student of his from 1965–70.

A special thank you goes to Nancy Thorndike Greenspan, author of *Atomic Spy: The Dark Lives of Klaus Fuchs,* who spent an hour on the phone with me one Sunday afternoon explaining why a rabbit hole I was pursuing would be pointless.

Finally, a very special thank you to Dan Mugg, my partner of twenty years, for putting up with all the things anyone involved with a writer has to put up with, including the long hours of my being physically here but mentally elsewhere and books all over the house. In the words of Sammy Cahn, if that isn't love, it will have to do until the real thing comes along.

About Me

I grew up in Oklahoma City and graduated from Central State College (now the University of Central Oklahoma) in 1970. My major was history. I had planned to teach, but I happened to graduate at a time when there was a teacher glut. I needed a job, so I accepted one with the federal government in procurement and contracting. I eventually left the government and did the same job for a major defense contractor. In the early 1990s, I was influenced by the writings of Joe Dominguez and Vicki Robin (*Your Money or Your Life,* 1992) and Paul Terhorst (*Cashing in on the American Dream,* 1988) and did some writing in what was then called the New Frugality Movement. The same principles are now espoused by what is called FIRE (financial independence, retire early). In 1995 I was able to leave my job, move to where I wanted to live, and pursue my interests. My 1995 book, *Beating the System,* tells that story. Other books I wrote and edited were published by NTC Contemporary Books and Berkley. All are available cheap on Amazon. Things worked out well, and one day I may write a sequel, perhaps titled *Still Beating the System After All These Years.* I accepted a consulting gig in 1997 that lasted four years.

I've always been interested in history, and a few years ago I learned about a program in Missouri that lets senior citizens audit classes on a space available basis. Other states, I'm sure, have similar programs. I have been auditing classes for several years at the University of Missouri-Kansas City. I've come to know several excellent professors and have learned at least as much as I did the first time I went to college. And I've learned that I had to unlearn some of the things I learned the first time around.

I very occasionally post to my blog: larryrothsblog.com.

Thanks for reading.

Conferences
Policy Studies Organization hosts numerous conferences, including the Middle East Dialogue, Space Education and Strategic Applications, International Criminology Conference, Dupont Summit on Science, Technology and Environmental Policy, World Conference on Fraternalism, Freemasonry and History, AI – The Future of Education: Disruptive Teaching and Learning Models, Sport Management and Esport Conference, and the Internet Policy & Politics Conference. Recordings of these talks are available in the PSO Video Library.

Yearbook
The Policy Yearbook contains a detailed international listing of policy scholars with contact information, fields of specialization, research references, and an individual scholar's statements of research interests.

Curriculum Project
The Policy Studies Organization aims to provide resources for educators, policy makers, and community members, to promote the discussion and study of the various policies that affect our local and global society. Our curriculum project organizes PSO articles and other media by easily serachable themes.

For more information on these projects, access videos of past talks, and upcoming events, please visit us at:

ipsonet.org

Related Titles from Westphalia Press

The Athenian Year Primer: Attic Time-Reckoning and the Julian Calendar
by Christopher Planeaux

The ability to translate ancient Athenian calendar references into precise Julian-Gregorian dates will not only assist Ancient Historians and Classicists to date numerous historical events with much greater accuracy but also aid epigraphists in the restorations of numerous Attic inscriptions.

Siddhartha: Life of the Buddha
by David L. Phillips,
contributions by Venerable Sitagu Sayadaw

Siddhartha: Life of the Buddha is an illustrated story for adults and children about the Buddha's birth, enlightenment and work for social justice. It includes illustrations from Pagan, Burma which are provided by Rev. Sitagu Sayadaw.

Growing Inequality: Bridging Complex Systems, Population Health, and Health Disparities
Editors: George A. Kaplan, Ana V. Diez Roux, Carl P. Simon, and Sandro Galea

Why is America's health is poorer than the health of other wealthy countries and why health inequities persist despite our efforts? In this book, researchers report on groundbreaking insights to simulate how these determinants come together to produce levels of population health and disparities and test new solutions.

Issues in Maritime Cyber Security
Edited by Dr. Joe DiRenzo III, Dr. Nicole K. Drumhiller, and Dr. Fred S. Roberts

The complexity of making MTS safe from cyber attack is daunting and the need for all stakeholders in both government (at all levels) and private industry to be involved in cyber security is more significant than ever as the use of the MTS continues to grow.

Bunker Diplomacy: An Arab-American in the U.S. Foreign Service
by Nabeel Khoury

After twenty-five years in the Foreign Service, Dr. Nabeel A. Khoury retired from the U.S. Department of State in 2013 with the rank of Minister Counselor. In his last overseas posting, Khoury served as deputy chief of mission at the U.S. embassy in Yemen (2004-2007).

Managing Challenges for the Flint Water Crisis
Edited by Toyna E. Thornton, Andrew D. Williams, Katherine M. Simon, Jennifer F. Sklarew

This edited volume examines several public management and intergovernmental failures, with particular attention on social, political, and financial impacts. Understanding disaster meaning, even causality, is essential to the problem-solving process.

User-Centric Design
by Dr. Diane Stottlemyer

User-centric strategy can improve by using tools to manage performance using specific techniques. User-centric design is based on and centered around the users. They are an essential part of the design process and should have a say in what they want and need from the application based on behavior and performance.

Masonic Myths and Legends
by Pierre Mollier

Freemasonry is one of the few organizations whose teaching method is still based on symbols. It presents these symbols by inserting them into legends that are told to its members in initiation ceremonies. But its history itself has also given rise to a whole mythology.

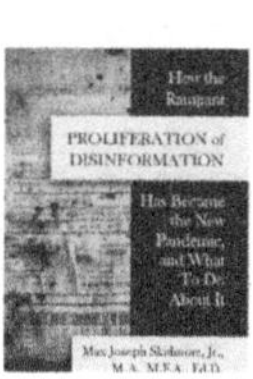

How the Rampant Proliferation of Disinformation has Become the New Pandemic by Max Joseph Skidmore Jr.

This work examines the causes of the overwhelming tidal wave of fake news, misinformation, disinformation, and propaganda, and the increase in information illiteracy and mistrust in higher education and traditional, vetted news outlets that make fact-checking a priority

Anti-Poverty Measures in America: Scientism and Other Obstacles
Editors, Max J. Skidmore and Biko Koenig

Anti-Poverty Measures in America brings together a remarkable collection of essays dealing with the inhibiting effects of scientism, an over-dependence on scientific methodology that is prevalent in the social sciences, and other obstacles to anti-poverty legislation.

Geopolitics of Outer Space: Global Security and Development
by Ilayda Aydin

A desire for increased security and rapid development is driving nation-states to engage in an intensifying competition for the unique assets of space. This book analyses the Chinese-American space discourse from the lenses of international relations theory, history and political psychology to explore these questions.

Contests of Initiative: Countering China's Gray Zone Strategy in the East and South China Seas
by Dr. Raymond Kuo

China is engaged in a widespread assertion of sovereignty in the South and East China Seas. It employs a "gray zone" strategy: using coercive but sub-conventional military power to drive off challengers and prevent escalation, while simultaneously seizing territory and asserting maritime control.

Discourse of the Inquisitive
Editors: Jaclyn Maria Fowler and Bjorn Mercer

Good communication skills are necessary for articulating learning, especially in online classrooms. It is often through writing that learners demonstrate their ability to analyze and synthesize the new concepts presented in the classroom.

westphaliapress.org

Made in the USA
Monee, IL
13 September 2026

62975275R00125